TC 3-23.35

Pistol

MAY 2017

DISTRIBUTION RESTRICTION: Approved for public release; distribution is unlimited.
This publication supersedes FM 3-23.35 (Combat Training with Pistols, M9 and M11), 25 June 2003.

Headquarters, Department of the Army

*TC 3-23.35 (FM 3-23.35)

Training Circular
No. 3-23.35 (FM 3-23.35)

Headquarters
Department of the Army
Washington, DC, 30 May 2017

Pistol

Contents

Page

PREFACE v

Chapter 1	OVERVIEW 1-1	
	Safe Weapons Handling 1-2	
	Rules of Firearms Safety 1-3	
	Weapon Safety Status 1-4	
	Weapon Control Status 1-5	
Chapter 2	PRINCIPLES OF OPERATION 2-1	
	Description 2-1	
	Major Components 2-2	
	Cycle of Function 2-5	
Chapter 3	AIMING DEVICES 3-1	
	Iron Sight 3-1	
	AN/PEQ-14 3-2	
	Operator Controls 3-6	
Chapter 4	HOLSTERS AND ACCESSORIES 4-1	
	Requirements 4-1	
	Placement 4-2	
	Accessories 4-4	
Chapter 5	EMPLOYMENT 5-1	
	Firing Situations 5-1	
	Shot Process 5-1	

Distribution Restriction: Approved for public release; distribution is unlimited.

*This publication supersedes FM 3-23.35 (Combat Training with Pistols, M9 and M11), 25 June 2003.

Contents

	Target Acquisition 5-3
	Draw and Present the Pistol 5-5
	Disengage the Safety 5-11
	Reholster the Pistol 5-11
Chapter 6	**STABILITY** **6-1**
	Support 6-1
	Muscle Relaxation 6-2
	Natural Point of Aim 6-2
	Recoil Management 6-3
	Aspects of Grip 6-3
	One- and Two-Hand Grips 6-7
	Shooter-Gun Angle 6-9
	Field of View 6-9
	Carry Positions 6-9
	Stabilization 6-10
	Firing Positions 6-11
Chapter 7	**AIM** **7-1**
	Elements of Accuracy 7-1
	Common Engagements 7-2
	Common Aiming Errors 7-6
Chapter 8	**CONTROL** **8-1**
	Arc of Movement 8-1
	Malfunctions 8-6
Chapter 9	**MOVEMENT** **9-1**
	Movement Techniques 9-1
	Forward Movement 9-2
	Retrograde Movement 9-3
	Lateral Movement 9-3
	Turning Movement 9-4
Appendix A	**AMMUNITION** **A-1**
Appendix B	**BALLISTICS** **B-1**
Appendix C	**COMPLEX ENGAGEMENTS** **C-1**
Appendix D	**DRILLS** **D-1**
Appendix E	**QUALIFICATION** **E-1**
	GLOSSARY **Glossary-1**
	REFERENCES **References-1**
	INDEX **Index-1**

Contents

Figures

Figure 1-1. Employment skills 1-1
Figure 2-1. M9 service pistol 2-1
Figure 2-2. Slide and barrel assembly 2-3
Figure 2-3. Receiver assembly 2-4
Figure 2-4. Feeding example 2-6
Figure 2-5. Chambering example 2-7
Figure 2-6. Locking example 2-8
Figure 2-7. Firing example 2-9
Figure 2-8. Unlocking example 2-10
Figure 2-9. Extracting example 2-11
Figure 2-10. Ejecting example 2-12
Figure 2-11. Cocking example 2-13
Figure 3-1. Front and rear sight 3-1
Figure 3-2. Three-dot sight system 3-2
Figure 3-3a. AN/PEQ-14 operator controls (front) 3-6
Figure 3-3b. AN/PEQ-14 operator controls (back) 3-7
Figure 3-4. Remote cable switch 3-9
Figure 3-5. Pattern generator install 3-12
Figure 4-1. Holsters 4-1
Figure 4-2. Placement 4-3
Figure 4-3. Weapon-mounted light 4-4
Figure 5-1a. Prepare to draw 5-6
Figure 5-1b. Grip and defeat 5-7
Figure 5-1c. Draw and rotate 5-8
Figure 5-1d. Meet and greet 5-9
Figure 5-1e. Extend and prepare 5-10
Figure 5-2. Methods for defeating the safety 5-11
Figure 6-1a. Force and counterforce in the pistol grip (one-hand grip) 6-4
Figure 6-1b. Force and counterforce in the pistol grip (two-hand grip) 6-4
Figure 6-2. Pistol leverage 6-6
Figure 6-3. One-handed grip (right-hand example) 6-7

Figure 6-4. Two-handed grip (right-hand example) 6-8
Figure 6-5. Carry positions 6-9
Figure 6-6. Firing stability 6-10
Figure 6-7. Standing unsupported position 6-11
Figure 6-8. Standing supported position 6-12
Figure 6-9. Kneeling unsupported position 6-13
Figure 6-10. Kneeling supported 6-14
Figure 6-11. Prone unsupported position 6-15
Figure 6-12. Prone supported position 6-16
Figure 7-1. Horizontal weapon orientation, example 7-3
Figure 7-2. Vertical weapon orientation, example 7-4
Figure 7-3. Proper sight alignment 7-5
Figure 8-1. Example of arc of movement 8-1
Figure 8-2. Trigger finger placement 8-3
Figure 8-3. Workspace management 8-4
Figure 9-1. Horizontal movements 9-2
Figure A-1. M882 ball round A-3
Figure B-1. Lethal zone example B-8
Figure C-1. Weapon mount and flashlight C-1
Figure C-2. Tracking method C-3
Figure C-3. Trapping method C-4
Figure E-1. Sample completed DA Form 88 .... E-10

Tables

Table 1-1. Weapon safety status for the service pistol 1-5
Table 1-2. Example of weapon control status 1-5
Table 3-1. AN/PEQ-14 modes of operation 3-3
Table 3-2. AN/PEQ-14 specifications 3-5
Table 4-1. Ordering information for M9 accessories 4-4
Table 5-1. Shot process example 5-2
Table E-1. Target-exposure times E-2

Preface

This publication provides the framework and techniques for conducting pistol training; the components and cycle of function for the M9 service pistol; its characteristics, equipment, and ammunition; and covers the functional elements of the shot process that are essential to build Soldier proficiency with their service pistol. It serves as an authoritative reference for personnel developing institutional, installation, and standard operating procedures (SOPs) for pistol training and qualification.

The principal audience for Training Circular (TC) 3-23.35 includes commanders, staff, leaders, Soldiers, trainers, and instructors who are responsible for planning, preparing, executing, and assessing pistol training and employment.

Doctrine is descriptive in nature but requires judgement in application. This publication outlines the appropriate use of the service pistol.

Commanders, staffs, and subordinates ensure that their decisions and actions comply with applicable United States, international, and in some cases host-nation laws and regulations. Commanders at all levels ensure that their Soldiers operate in accordance with the law of war and the rules of engagement. (See FM 27-10, *The Law of Land Warfare*.)

Joint terms are used as applicable. Selected joint and Army terms and definitions appear in both the glossary and the text. Terms for which this manual is proponent (author) are italicized in text and marked with an asterisk (*) in the glossary. Terms and definitions prescribed by this publication are boldfaced in text. Other terms defined in text are italicized and followed by the number of the proponent publication in parentheses.

TC 3-23.35 applies to the Active Army, Army National Guard/Army National Guard of the United States and United States Army Reserve unless otherwise stated.

The proponent for this publication is the Maneuver Center of Excellence (MCoE). Users and readers of this publication are invited to submit recommendations that will improve its effectiveness. You may send comments and recommendations following the format of DA Form 2028 *(Recommended Changes to Publications and Blank Forms)*. Point of contact information is:

E-mail:	usarmy.benning.mcoe.mbx.doctrine@mail.mil
Phone	COM 706-545-7114 or DSN 835-7114
Fax	COM 706-545-8511 or DSN 835-8511
U.S. Mail:	Commander, MCoE
	Directorate of Training and Doctrine (DOTD)
	Doctrine and Collective Training Division
	ATTN: ATZK-TDD
	Fort Benning, GA 31905-5410

This page intentionally left blank.

Introduction

This manual is comprised of nine chapters and five appendices, and is specifically tailored to the individual Soldier's use of the M9 service pistol. This TC provides specific information about the weapon, aiming devices, attachments, followed by sequential chapters on the tactical employment of the weapon system.

The training circular itself is purposely organized in a progressive manner, each chapter or appendix building on the information from the previous section. This organization provides a logical sequence of information which directly supports the Army's training strategy for the weapon at the individual level.

Chapters 1 through 4 describe safety, operation, types of sights, and accessories associated with the M9 service pistol. General information is provided in the chapters of the manual, with more advanced information placed in appendix A, Ammunition, and appendix B, Ballistics.

Chapters 5 through 9 provide the employment, stability, aiming, control and movement information. This portion focuses on the Solider skills needed to produce well aimed shots. Advanced engagement concepts are provided in appendix C of this publication. Appendix D of this publication provides common tactical drills that are used in training and combat that directly support tactical engagements. Appendix E of this publication provides information about qualificaton.

This manual does not cover the specific M9 service pistol training strategy, ammunition requirements for the training strategy, or range operations. These areas will be covered in separate training circulars.

Conclusion

TC 3-23.35 applies to all Soldiers, regardless of experience or position. This publication is designed specifically for the Soldier's use on the range during training, and as a reference while deployed.

This page intentionally left blank.

Chapter 1
Overview

This chapter describes the principles of safe weapon handling, provides the rules of firearm safety, and discusses weapon safety and control status.

Each Soldier must place accurate fires on threat targets with their individual weapon. To do this, the Soldier must understand the functional elements of the shot process, the principles of operation of the weapon, the characteristics and description of ammunition, and the various engagement techniques essential to building the Soldier's proficiency with their weapon. The combination of knowledge and practice, builds and sustains the skills to achieve accurate and precise shots consistently during combat operations. (See figure 1-1).

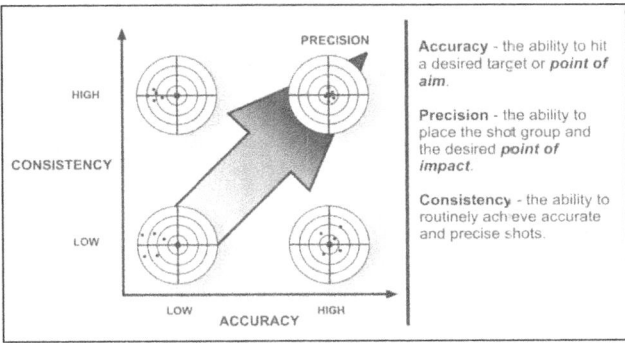

Figure 1-1. Employment skills

SAFE WEAPONS HANDLING

1-1. Safe weapons handling procedures are a consistent and standardized way for Soldiers to handle, operate, and employ the weapon safely and effectively. Weapons handling is built on three components; the Soldier, the weapon, and the environment, which are discussed below:

- The Soldier must maintain situational understanding of friendly forces, the status of the weapon, and the ability to evaluate the environment to properly handle any weapon. The smart, adaptive, and disciplined Soldier is the primary safety mechanism for all weapons under their control.
- The weapon is the primary tool of the Soldier to defeat threats in combat. The Soldier must know of and how to operate the mechanical safeties built into the weapons they employ, as well as the principles of operation for those weapons.
- The environment is the Soldier's surroundings. The Soldier must be aware of muzzle discipline, the nature of the target, and what is behind it.

1-2. Soldiers must be cognitively aware of three distinct weapons handling measures to safely and effectively handle weapons The weapons handling measures are:

- Rules of firearms safety.
- Weapons safety status.
- Weapons control status.

1-3. The weapons handling measures provide redundant safety measures when handling any weapon or weapon system in training and operational environments. A Soldier would have to violate two of the rules of firearms safety or violate a weapon safety status in order to have a negligent discharge.

Note. Unit standard operating procedures (SOPs), range SOPs, or the operational environment may dictate additional safety protocols; however, the rules of firearms safety are always applied. If a unit requires Soldiers to violate these safety rules for any reason, such as for the use of blank rounds or other similar training munitions during training, the unit commander must take appropriate risk mitigation actions.

Overview

RULES OF FIREARMS SAFETY

1-4. The rules of firearms safety are standardized for any weapon a Soldier may employ. Soldiers must adhere to these precepts during training and combat operations, regardless of the type of ammunition used.

RULE 1: TREAT EVERY WEAPON AS IF IT IS LOADED

1-5. Any weapon handled by a Soldier must be treated as if it is loaded and prepared to fire. Whether or not a weapon is loaded should not affect how a Soldier handles the weapon. Soldiers must take the appropriate actions to ensure the proper weapon status is applied during operations, whether in combat or training.

RULE 2: NEVER POINT THE WEAPON AT ANYTHING YOU DO NOT INTEND TO SHOOT

1-6. Soldiers must be aware of the orientation of their weapon's muzzle and what is in the path of the projectile if the weapon fires. Soldiers must ensure the path between the muzzle and target is clear of friendly forces, noncombatants, or anything the Soldier does not want to shoot.

1-7. When this is unavoidable, the Soldier must minimize the amount of time the muzzle is oriented toward people or objects they do not intend to shoot, while simultaneously applying the other three rules of fire arms safety.

RULE 3: KEEP FINGER STRAIGHT AND OFF THE TRIGGER UNTIL READY TO FIRE

1-8. Soldiers must not place their finger on the trigger unless they intend to fire the weapon. The Soldier is the most important safety feature on any weapon. Mechanical safety devices are not available on all types of weapons. When mechanical safeties are present, Soldiers must not solely rely upon them for safe operation knowing that mechanical measures may fail.

1-9. Whenever possible, Soldiers should move the weapon to mechanical safe when a target is not present. If the weapon does not have a traditional mechanical safe, the trigger finger acts as the primary safety.

RULE 4: ENSURE POSITIVE IDENTIFICATION OF THE TARGET AND ITS SURROUNDINGS

1-10. The disciplined Soldier can positively identify the target and knows what is in front of and what is beyond it. The Soldier is responsible for all bullets fired from their weapon, including the projectile's final destination.

1-11. Application of this rule minimizes the possibility of fratricide, collateral damage, or damage to infrastructure or equipment. It also prepares the Soldier for any follow-on shots that may be required.

Chapter 1

WEAPON SAFETY STATUS

1-12. The readiness of a Soldier's weapon is termed its weapon safety status (WSS). It is standard code that uses common colors (green, amber, red, and black) to represent the level of readiness for a given weapon. Each color represents a specific series of actions applied to a weapon. They are used in training and combat to place or maintain a level of safety relevant to the current task or action of a Soldier, small unit, or group. Table 1-1 shows the weapon safety status for the Army service pistol.

Note. If the component, assembly, or part described are unclear, refer to technical manual (TM) 9-1005-317-10, *Operator Manual Pistol, Semiautomatic, 9mm, M9 (1005-01-118-2640) (EIC: 4mn) Pistol, Semiautomatic, 9mm, M9A1 (1005-01-525-7966) Pistol, Semiautomatic, 9mm, Go Pistol (1005-01-588-5964) Air Force Only Go Pistol (1005-01-480-1274)* or chapter 2 of this publication.

GREEN

1-13. The weapon's magazine is out, the chamber is empty, and its slide is locked open or forward, and the decoking or safety lever is on SAFE.

Note. The command given to direct a GREEN safety status is, CLEAR.

AMBER

1-14. A magazine is locked into the magazine well of the weapon, the slide is forward on an empty chamber, and the decocking or safety lever is on SAFE.

Note. The command given to direct an AMBER status is, LOAD MAGAZINE.

RED

1-15. The weapon's magazine is inserted, a round is in the chamber, the slide is forward and locked, and the decocking or safety lever is on SAFE. The M9 service pistol can be carried in the holster with the safety/decocking lever in the up (off) position. If the Soldier chooses to carry the pistol with the safety/decocking lever down (on), the Soldier can defeat the safety at various points during the draw. Leadership will decide on the safety/decocking status in the holster based on SOP.

Note. The command given to direct a RED safety status is, MAKE READY

Overview

BLACK

1-16. The weapon's magazine is inserted, a round is in the chamber, the slide is forward and locked, and the decocking or safety lever is on FIRE. The Soldier's finger is on the trigger. The Soldier has a clear path from the muzzle of the pistol to the target.

Note. The command given to direct a BLACK safety status is driven by the unit's SOP, rules of engagement (ROE), or the command, FIRE.

Table 1-1. Weapon safety status for the service pistol

STATUS	GREEN	AMBER	RED	BLACK
Function	Clear	Prepared	Ready, safe	Ready, fire
Commands	CLEAR	LOAD MAGAZINE	MAKE READY	SOP/ROE/ FIRE
Ammunition	None	Magazine in	Magazine in/ round chambered	Magazine in/ round chambered
Slide	Locked open or forward	Forward	Forward	Forward
Chamber	Empty	Empty	Locked	Locked
Safety	Safe	Safe	Safe	Fire
Trigger	Off	Off	Off	On

WEAPON CONTROL STATUS

1-17. When applicable, the leader may impose a weapon control status (WCS) in addition to the weapon safety status. The weapon control status outlines the conditions, based on target identification criteria, under which friendly elements may engage. This status is adjustable, as necessary, based on the current rules of engagement established for the area of operations.

1-18. Table 1-2 provides a description of the standard weapon control statuses used in conjunction with the weapon safety status. They describe when the Soldier is authorized to engage a threat target once the threat conditions have been met.

Table 1-2. Example of weapon control status

WEAPON CONTROL STATUS	DESCRIPTION
Weapon hold	Engage only if engaged or ordered to engage.
Weapon tight	Engage only if target is positively identified as enemy.
Weapon free	Engage targets not positively identified as friendly.

This page intentionally left blank.

Chapter 2
Principles of Operation

This chapter provides the general characteristics, description, components, and principles of operation for the M9 service pistol. It provides a general overview of the mechanics and theory of how weapon operates, key terms and definitions related to their functioning, and the physical relationship between the Soldier and the weapon.

DESCRIPTION

2-1. The M9 service pistol is a semiautomatic, magazine-fed, recoil-operated, double- and single-action weapon that is chambered for the 9- by 19-mm (millimeter) NATO cartridge. (See figure 2-1.) (For more information about the technical aspects of the M9 service pistol, refer to TM 9-1005-317-10.)

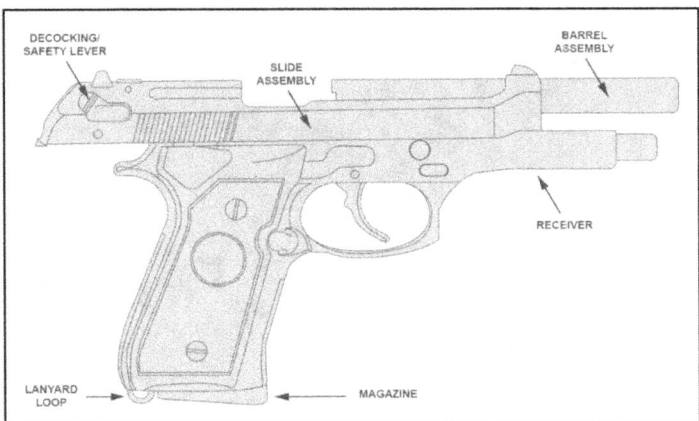

Figure 2-1. M9 service pistol

2-2. The M9 service pistol incorporates single- and double-action modes of fire. For double action, pulling the trigger will cock the hammer and immediately release it, discharging the first chambered round. To fire the first chambered round in single action,

Chapter 2

the Soldier manually locks the hammer to the rear (fully cocked position) before pulling the trigger. The Soldier fires all shots after the first one in single action, because the slide automatically cocks the hammer after each shot.

2-3. The M9 service pistol has a short recoil system using a falling locking block. Upon firing, the pressure developed by the combustion gases recoils the slide/barrel assembly. After a short run, the locking block will stop the rearward movement of the barrel and release the slide, which will continue its rearward movement. The slide will then extract and eject the case, cock the hammer and compress the recoil spring. Then the slide moves forward, feeding the next round from the magazine into the chamber. The slide should remain locked to the rear and the chamber open after the last round has been fired and ejected if the magazine is present.

2-4. The M9 service pistol consists of components, assemblies, subassemblies, and individual parts. Soldiers must be familiar with these items and how they interact during operation:

- Components are a uniquely identifiable group of fitted parts, pieces, assemblies or subassemblies that are required and necessary to perform a distinctive function in the operation of the weapon. Components are usually removable in one piece and are considered indivisible for a particular purpose or use.
- Assemblies are a group of subassemblies and parts that are fitted to perform specific set of functions during operation, and cannot be used independently for any other purpose.
- Subassemblies are a group of fitted parts that perform a specific set of functions during operation. Subassemblies are compartmentalized to complete a specific task. They may be grouped with other assemblies, subassemblies and parts to create a component.
- Parts are the individual items that perform a function when attached to a subassembly, assembly, or component that serves a specific purpose.

MAJOR COMPONENTS

2-5. The M9 service pistol has two major components: the slide assembly and the receiver. These components are described below including their associated assemblies, subassemblies, and parts.

Principles of Operation

SLIDE ASSEMBLY

2-6. An open top steel slide with a Bruniton finish to protect the steel parts from rain and corrosion. The open slide design aids in eliminating jamming and stove piping. It also allows the Soldier to load and chamber one cartridge at a time should the magazine be lost or damaged. Figure 2-2 describes each component of the slide and barrel assembly.

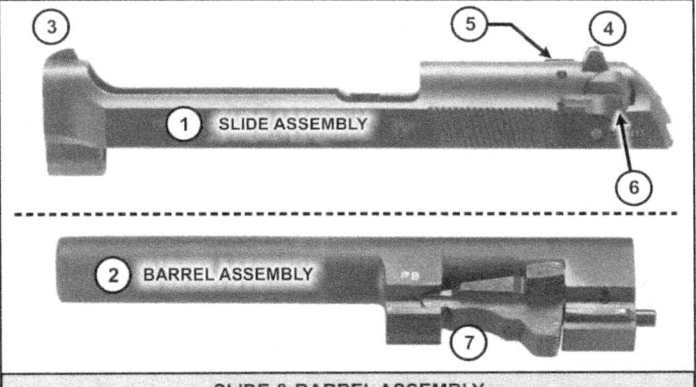

SLIDE & BARREL ASSEMBLY

1. Slide assembly. Houses the firing pin, firing pin striker, and extractor. It cocks the hammer during the recoil cycle.
2. Barrel assembly. Houses the cartridge for firing, directs the projectile, and locks the barrel in position during firing.
3. Front sight. It is a blade design that is integrated with the slide.
4. Rear sight. The rear sight is designed to provide a front projection so that in an emergency, the Soldier may retract the slide single-handedly by pushing the rear sight against the edge of a table, door, or any other hard structure.
5. Firing pin block. The front part of the firing pin is blocked from any forward movement until the trigger is pulled completely back. The block is located rearward, far away from the fouling and debris of the breech face. Since the block is visible, the user may ascertain its proper operation at any time.
6. Decocking/safety lever. Accommodates both left- and right-handed Soldiers. Decocks the pistol without causing an accidental discharge and has two positions, Fire (the up position) and Safe (the down position).
7. Locking block. The M9 9-mm pistol uses a pivoting block design that locks the barrel into place.

Figure 2-2. Slide and barrel assembly

Chapter 2

Receiver

2-7. The receiver consists of the following components, assemblies, and parts as seen in figure 2-3.

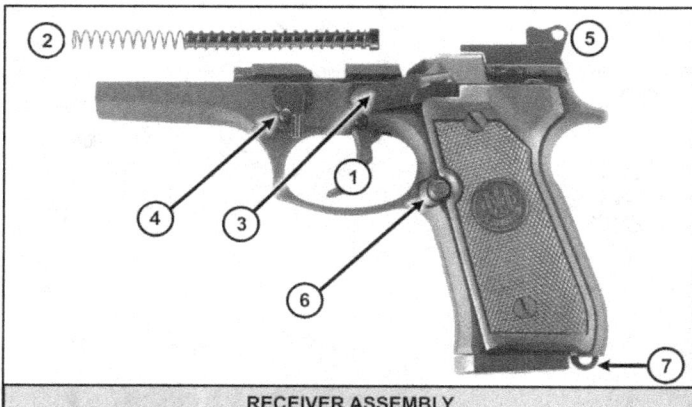

RECEIVER ASSEMBLY

1. Trigger assembly. Consists of the trigger, trigger pins, springs, and other mechanical components necessary to fire the weapon.
2. Recoil spring and recoil spring guide. Absorbs recoil and returns the slide assembly to the forward position after firing the weapon.
3. Slide stop. Holds the slide to the rear after the last cartridge has been fired. It can also be engaged manually.
4. Disassembly lever and disassembly buttons. Allows quick field stripping. Prevents accidental disassembly.
5. Hammer. Has three positions: decocked, half-cocked, or fully cocked.
6. Magazine catch assembly. Provides a spring loaded locking mechanism to secure the magazine within the magazine well. Provides the Soldier an easy-to-manipulate, push-to-release textured button that releases the magazine from the magazine well during operation. The magazine catch assembly is reversible for left-handed Soldiers.
7. Lanyard loop. Compatible with standard lanyards.

Figure 2-3. Receiver assembly

Principles of Operation

CYCLE OF FUNCTION

2-8. The cycle of function is the mechanical process a weapon follows during operation. The information provided below is specific to the cycle of function as it pertains specifically to the M9 service pistol. (Appendix A covers the types of ammunition that may be used with the service pistol.)

2-9. The cycle starts when the pistol is ready with the slide locked to the rear, the chamber is clear, and a magazine inserted into the magazine well with at least one cartridge. From this state, the cycle executes the sequential phases of the cycle of functioning to fire a round and prepare the weapon for the next round. The phases of the cycle of function in order are—

- Feeding.
- Chambering.
- Locking.
- Firing.
- Unlocking.
- Extracting.
- Ejecting.
- Cocking.

2-10. For the weapon to operate correctly, semiautomatic weapons require a system of operation to complete the cycle of functioning. The M9 service pistol has a short recoil system utilizing a falling locking block. Upon firing, the pressure created by the combustion gases of the fired cartridge recoils the slide-barrel assembly. After a short run, the locking block stops reward movement of the barrel and releases the slide that continues its rearward movement. The slide extracts then ejects the fired cartridge case, cocks the hammer, and compresses the recoil spring. The slide moves forward, feeding the next cartridge, from the top of the magazine into the chamber.

Chapter 2

FEEDING

2-11. Feeding is the process of mechanically providing a cartridge of ammunition to the entrance of the chamber. (See figure 2-4.)

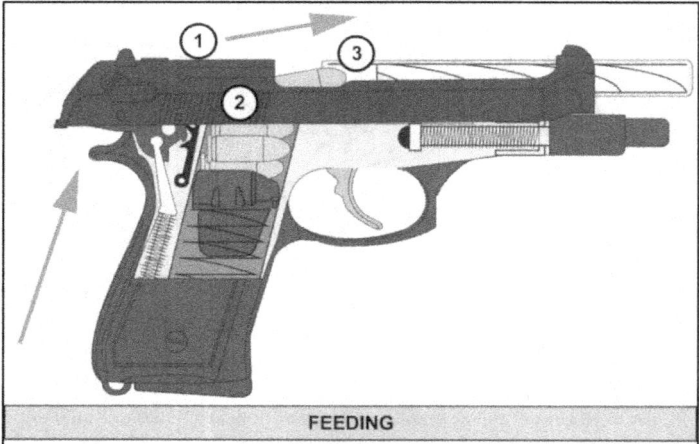

Figure 2-4. Feeding example

Principles of Operation

CHAMBERING

2-12. The chambering phase is the continuing action of feeding the round into the chamber of the weapon. (See figure 2-5.)

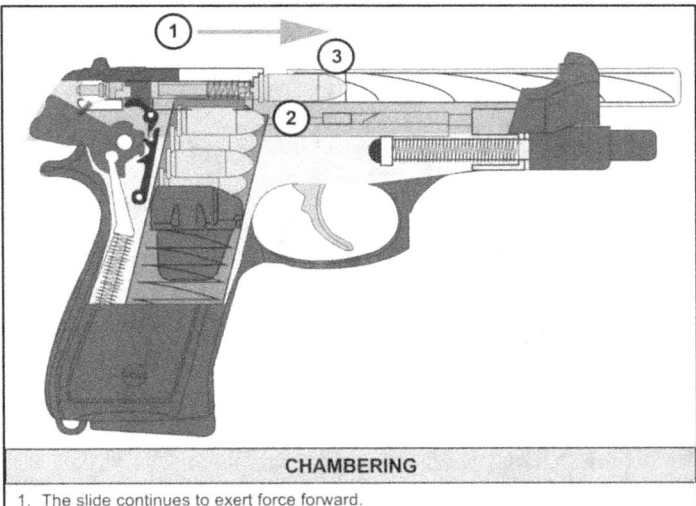

CHAMBERING
1. The slide continues to exert force forward.
2. The cartridge rides against a smooth inclined surface (chambering ramp) of the entrance to the chamber area.
3. The cartridge is driven up the feed ramp and enters the chamber, moving to the fully forward position. Chambering is completed when the shoulder or the cartridge makes contact with the corresponding shoulder area of the chamber.

Figure 2-5. Chambering example

Chapter 2

LOCKING

2-13. Locking is the process of creating a mechanical grip between the slide and chamber with the appropriate amount of headspace (clearance) for safe firing. (See figure 2-6.)

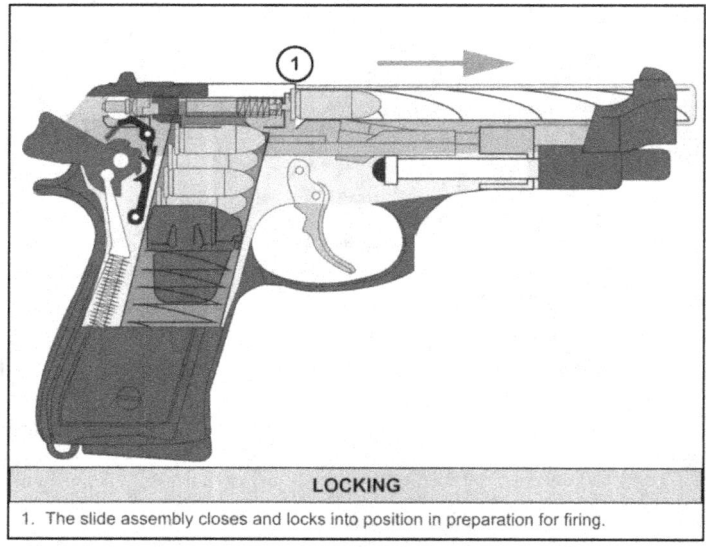

Figure 2-6. Locking example

Principles of Operation

FIRING

2-14. Firing is the finite process of initiating the primer detonation of the cartridge and continues through the shot exit of the projectile from the muzzle. (See figure 2-7.)

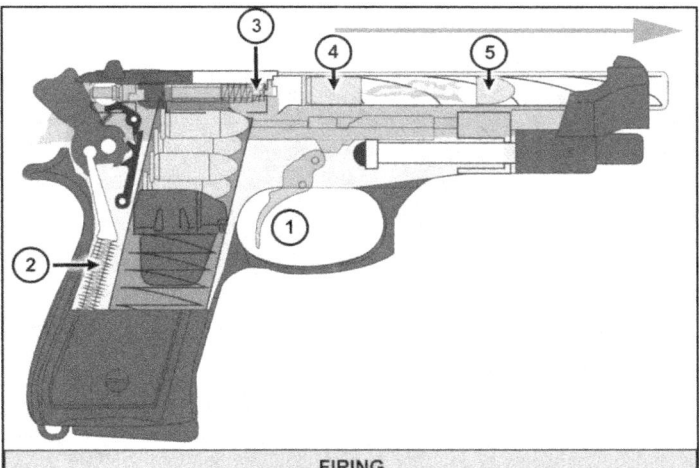

FIRING
1. Firing is initiated when the weapon is placed on fire, a round is chambered, and the trigger is squeezed.
2. The hammer spring exerts its force rotating the hammer forward, striking the firing pin.
3. The firing pin exerts its force in an equal and opposite direction, striking and igniting the primer on the rear center of the cartridge case.
4. Once the primer's charge is ignited, the round's propellant begins to burn. As the propellant burns, the cartridge case expands to the fullest extent of the chamber area, sealing the gases within the bore.
5. The expanding gas propels the projectile down the length of the bore.

Figure 2-7. Firing example

Chapter 2

UNLOCKING

2-15. Unlocking is the process of the slide beginning to come back. (See figure 2-8.)

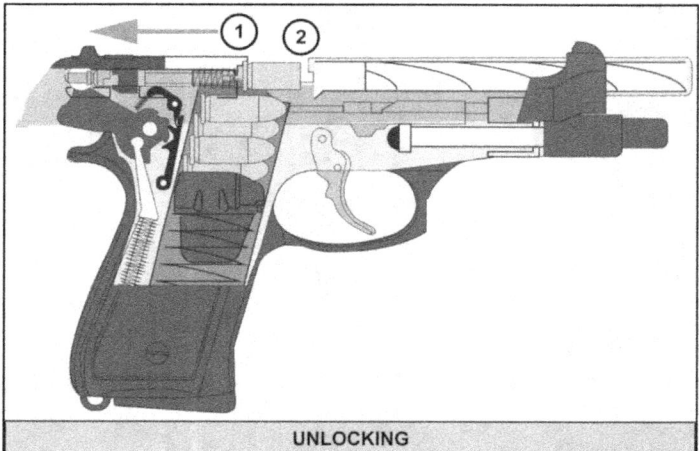

UNLOCKING
1. Once the round has fired, the spring tension and weight of the slide delays the slide moving rearward long enough for the chamber pressure to drop.
2. The barrel and slide move a small distance together and unlocking is complete at the end of this movement.

Figure 2-8. Unlocking example

Principles of Operation

EXTRACTING

2-16. Extracting is the removal of the expended cartridge case from the chamber by means of the extractor. (See figure 2-9.)

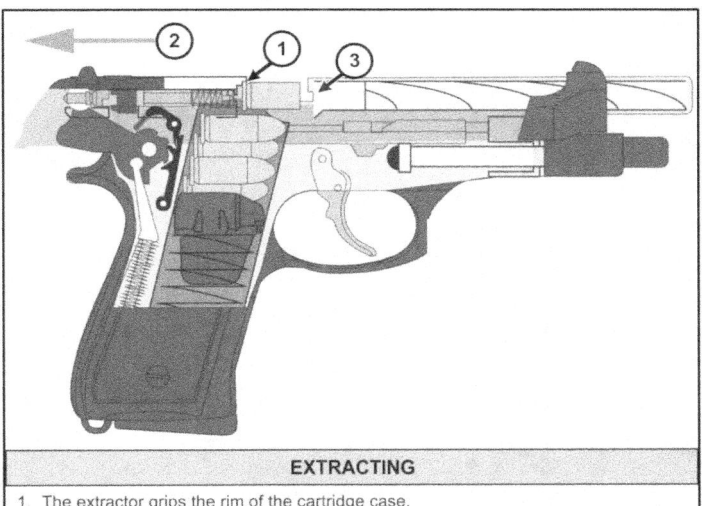

EXTRACTING
1. The extractor grips the rim of the cartridge case.
2. As the slide moves to the rear, the extractor pulls the expended cartridge case from the chamber.
3. The extracting phase continues until the expended cartridge case is clear of the chamber area but has not yet exited the weapon.

Figure 2-9. Extracting example

Chapter 2

Ejecting

2-17. Ejecting is the removal of the spent cartridge case from the weapon itself. (See figure 2-10.)

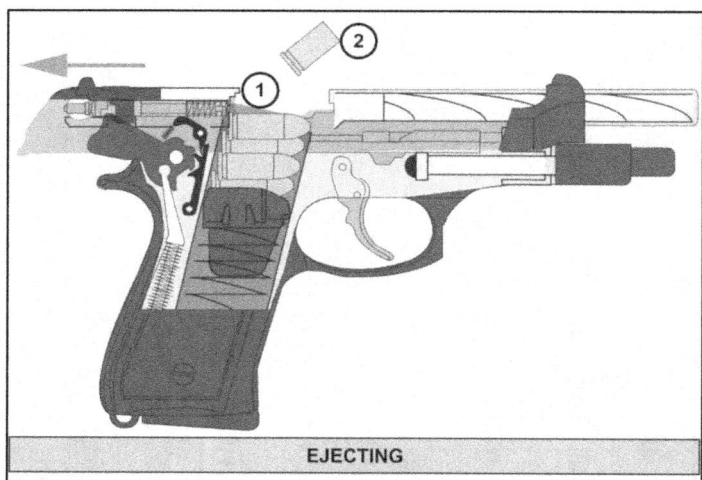

EJECTING
1. A rod, called the ejector rod, is housed in the frame assembly. This applies pressure against the left side of the spent cartridge case base.
2. The spent cartridge is ejected from the weapon.

Figure 2-10. Ejecting example

Principles of Operation

COCKING

2-18. Cocking is the process of mechanically positioning the trigger assembly for firing. (See figure 2-11.) The cocking phase completes the full cycle of functioning.

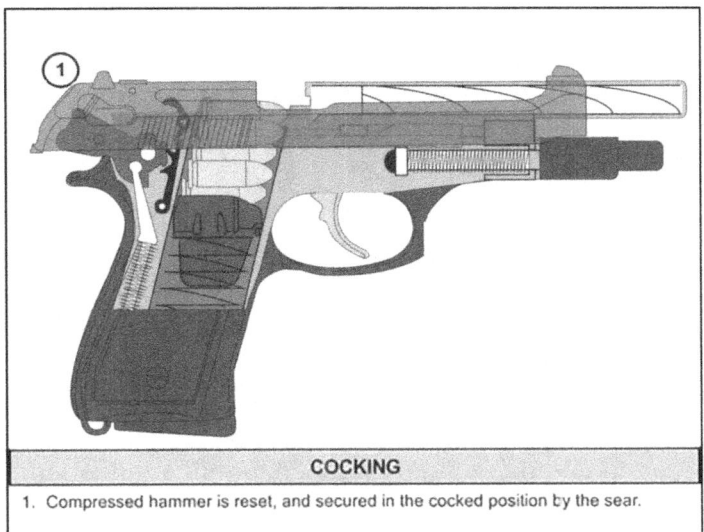

COCKING
1. Compressed hammer is reset, and secured in the cocked position by the sear.

Figure 2-11. Cocking example

This page intentionally left blank.

Chapter 3
Aiming Devices

This chapter provides the characteristics of the pistol iron sights and the attachable pointer/illuminator/lasers and their capabilities, function and use.

IRON SIGHT

3-1. The iron sight for the M9 service pistol is a fixed sighting system that includes a front sight blade and a rear notch. (See figure 3-1.)

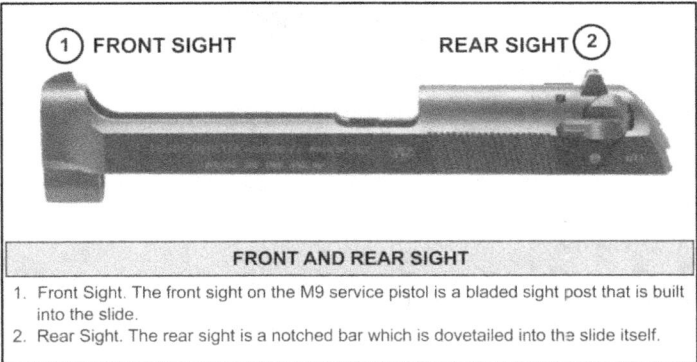

Figure 3-1. Front and rear sight

Chapter 3

3-2. The Army service pistol has a three-dot sight system embedded with the front sight and rear notch. When aiming, Soldiers should have even height and light between the front sight post and rear notch. Soldiers will refrain from using the dots to align the sights. (See figure 3-2.)

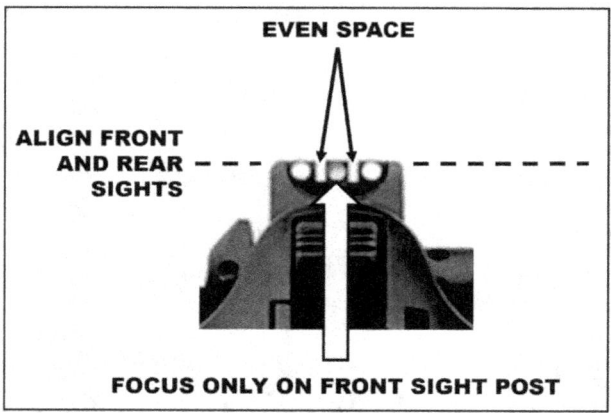

Figure 3-2. Three-dot sight system

AN/PEQ-14

3-3. The AN/PEQ-14 is a compact, lightweight device designed to be mounted on an M9 service pistol equipped with a tactical rail adapter. The AN/PEQ-14 contains an infrared (IR) aiming laser, IR illumination laser, visible red dot aiming laser and a white light (flashlight) with an adjustable focus integrated into one device. The visible and IR aiming lasers are co-aligned allowing both lasers to be boresighted simultaneously during daylight hours without the need of night vision devices (NVDs).

Aiming Devices

MODES OF OPERATION

3-4. To prevent inadvertent activation of the laser(s) or white light energy, the mode selector should be in the OFF position when not in use. For laser safety considerations, review TM 9-5855-1911-13&P, *Operator and Field Maintenance Manual Including Repair Parts and Special Tools for Integrated White Laser Pointer (ILWLP), AN/PEQ-14 (Black) NSN 0855-01-538-0191 (Tan) NSN 5855-01-571-1258*. (See table 3-1 for modes of operation.)

Table 3-1. AN/PEQ-14 modes of operation

KNOB POSITION	OPERATION	REMARKS
OFF	AN/PEQ-14 will not operate.	OFF prevents inadvertently turning the unit ON.
VIS	Visible aiming laser operates.	Visible aiming laser is enabled. Moving the toggle switch activates the laser.
VIS/ILL	Visible aiming laser operates with the visible illuminator.	The visible aiming laser and the visible white light illuminator are enabled. Moving the toggle switch will activate the laser and white light. Turning the laser inactivation switch enables white light only or VIS and ILL.
IR	Infrared aiming laser operates.	Infrared aiming laser is enabled. Moving the toggle switch activates the laser.
IR/ILL	IR aiming laser operates with the IR illuminator.	The IR aiming laser and the IR illuminator are enabled. Moving the toggle switch activates the laser and illuminator. Turing the laser inactivation switch enables the illuminator only or IR and ILL.
LEGEND	ILL illumination IR infrared VIS visible	

SPECIFICATIONS

3-5. Table 3-2 shows the AN/PEQ-14 mounted on an M9 pistol and provides a quick technical reference. (For more information about the AN/PEQ-14, refer to TM 9-5855-1911-13&P.)

Aiming Devices

Table 3-2. AN/PEQ-14 specifications

	TM 9-5855-1911-13&P		
	DIMENSIONS		
	LENGTH	4.1 in	10.4 cm
	WIDTH	1.9 in	4.8 cm
	HEIGHT	2.1 in	5.3 cm
	WEIGHT	6.0 oz	170 g
POWER			
BATTERY LIFE	30 minutes continuous white illumination		
POWER SOURCE	2 3-volt DL 123A		
MODE OF OPERATION			
KNOB POS	IR AIM	IR ILL	VIS AIM
OFF	OFF	OFF	OFF
VIS	OFF	OFF	ON
VIS/ILL	OFF	OFF	ON/WHITE LIGHT
IR	ON	OFF	OFF
IR/ILL	ON	ON	OFF
WHITE LIGHT	OFF	OFF	OFF
LASER	DIVERGENCE		WAVE LENGTH
IR AIM LASER	0.5 mRad		830 Nm
IR ILLUMINATOR	50 mRad		830 Nm
VIS LASER	0.5 mRad		650 Nm
LEGEND			
mRad – milliradians Nm – nanometers IR – infrared	in – inches oz – inches ILL – illumination		cm – centimeters g – grams VIS – visible POS – position

Chapter 3

OPERATOR CONTROLS

3-6. The AN/PEQ-14 has IR as well as visible white aiming light. Figures 3-3a and 3-3b identify the features and controls for the AN/PEQ-14.

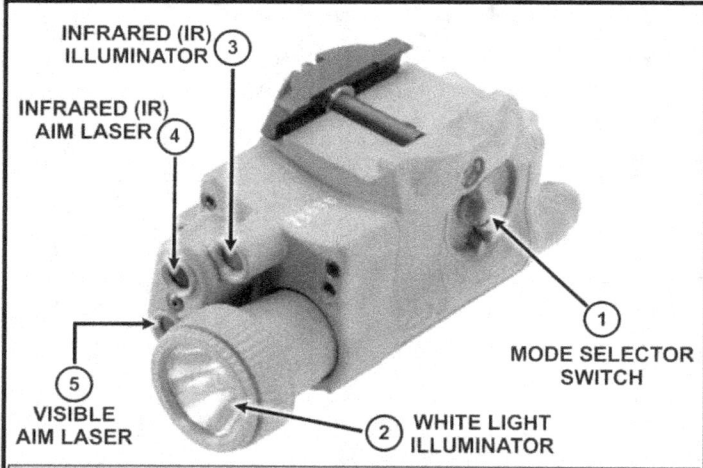

AN/PEQ-14 ILWLP (FRONTSIDE)

1. The mode selector switch has five positions: OFF, VIS, VIS/ILL, IR, and IR/ILL.
2. The white light illuminator is adjustable, narrow to wide, and will illuminate the target area in conditions of darkness.
3. The IR illuminator is used to illuminate the target area in conditions of darkness when using NVDs.
4. The IR aim laser is used to point at objects and targets during hours of darkness when using NVDs.
5. The visible aim laser is used to point at objects and targets during daylight or night conditions without the need for NVDs.

LEGEND		
	ILL	illumination
	ILWLP	integrated laser white light pointer
	IR	infrared
	NVD	night vision device
	VIS	visible

Figure 3-3a. AN/PEQ-14 operator controls (front)

Aiming Devices

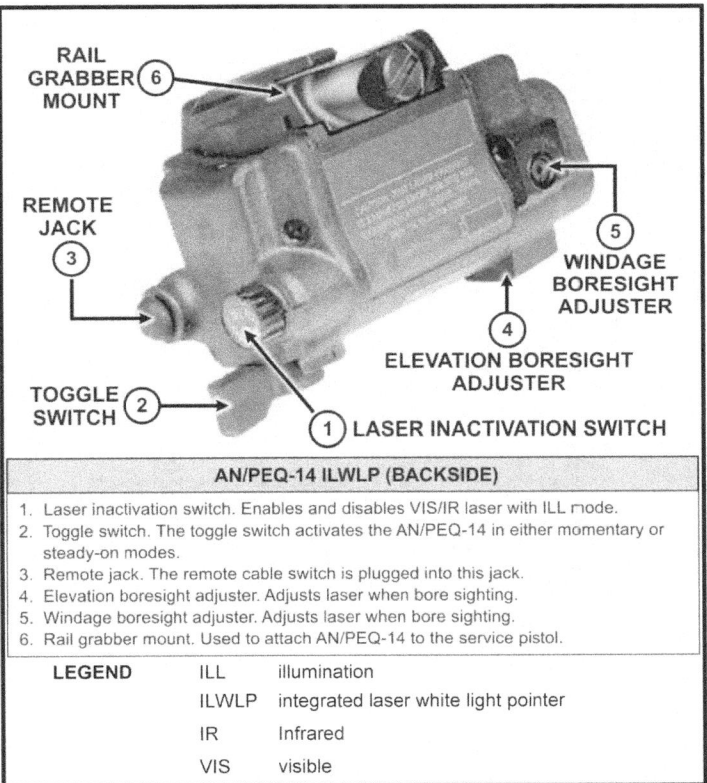

Figure 3-3b. AN/PEQ-14 operator controls (back)

AN/PEQ-14 ILWLP (BACKSIDE)

1. Laser inactivation switch. Enables and disables VIS/IR laser with ILL mode.
2. Toggle switch. The toggle switch activates the AN/PEQ-14 in either momentary or steady-on modes.
3. Remote jack. The remote cable switch is plugged into this jack.
4. Elevation boresight adjuster. Adjusts laser when bore sighting.
5. Windage boresight adjuster. Adjusts laser when bore sighting.
6. Rail grabber mount. Used to attach AN/PEQ-14 to the service pistol.

LEGEND
ILL — illumination
ILWLP — integrated laser white light pointer
IR — Infrared
VIS — visible

LASER ACTIVATION SWITCH

3-7. The laser activation switch enables deactivation of the visible aiming laser and allows operation of the white light illuminator only in the visible/illumination (VIS/ILL) mode. In addition, use of this switch allows deactivation of the infrared aiming laser and allows operation of the infrared illuminator only in the IR/ILL mode.

3-8. In order to achieve white light illuminator only and infrared illuminator only, use of the laser activation switch is required. Rotate the laser activation switch counterclockwise to enable the AN/PEQ-14 lasers to be turned on, and rotating clockwise turns the aiming lasers off.

TOGGLE SWITCH

3-9. The toggle switch activates the AN/PEQ-14 in either momentary or steady-on modes. For momentary operation, toggle the switch slightly to the left or right until the AN/PEQ-14 turns on. When the toggle is released, the AN/PEQ-14 turns off. For continuous operation, toggle the switch left or right until the toggle locks in a detent. When operating the AN/PEQ-14 in continuous mode, the operator must toggle the switch lever in the opposite direction until the detent is disengaged to turn the AN/PEQ-14 lasers or white light off.

REMOTE CABLE SWITCH

3-10. The AN/PEQ-14 comes with a jack plug installed in the remote jack that must be removed and stored before installing the remote cable switch. The remote cable switch (see figure 3-4) length is optimum for use with the AN/PEQ-14 on the M9 pistol equipped with a tactical rail adapter. The remote cable switch contains a pressure pad.

3-11. Pressing anywhere on the pressure pad activates the AN/PEQ-14 in the mode that is selected for as long as pressure is applied. The remote provides tactile feel and silent operation, and is secured to the weapon using hook and loop fastener tape to best suit the operator's firing preference.

Aiming Devices

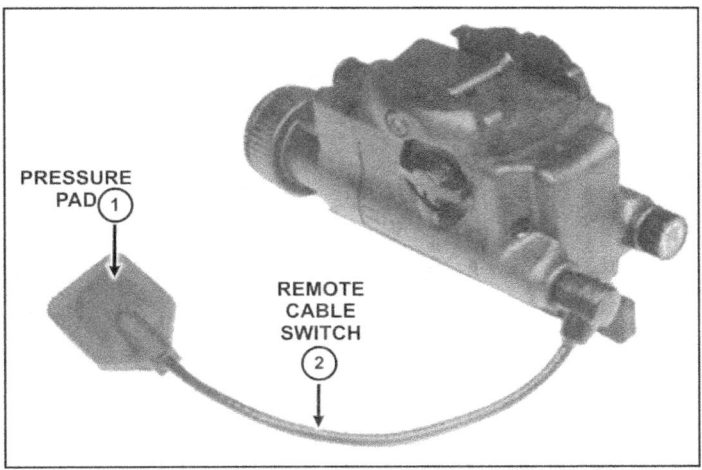

Figure 3-4. Remote cable switch

INFRARED ILLUMINATOR

3-12. The IR illuminator is used to illuminate the target area in conditions of darkness when using NVDs. The beam size is fixed to allow for illumination of a man-size target at 25 meters. The IR illuminator is aligned with the bore of the weapon at the factory.

3-13. The IR illuminator is a Class 3B device. The laser can reach a maximum of 100 meters in clear, quarter moon conditions. To operate the IR illuminator, turn the mode selector switch to IR/ILL and activate the lasers using the toggle switch or remote cable switch. To operate continuously, move the toggle switch to the STEADY-ON position or engage by pressing remote cable switch pressure pad.

3-14. To discontinue use, move the toggle switch to the OFF position or release pressure from the remote cable switch pressure pad. In the IR/ILL mode, the IR illuminator will operate in combination with the IR aiming laser. To deactivate the IR aiming laser, and operate in IR illuminator only, rotate the laser inactivation switch to the OFF position. Activate the IR illuminator using the toggle switch or remote cable switch.

Note. The infrared illuminator diffuser looks similar to the pattern generator and is installed using the same procedures.

3-15. The infrared illuminator diffuser diffuses (spreads) the laser energy over an angle approaching 180 degrees in front of the AN/PEQ-14. The infrared illuminator diffuser may be installed over the infrared illuminator to allow for illumination of a 3200-square foot, darkened enclosure. The infrared illuminator diffuser is contained within a shroud and is secured to the integrated laser white light pointer (ILWLP) body using same method as the pattern generators.

INFRARED AIM LASER

3-16. The IR aim laser is used to point at objects and targets during hours of darkness when using NVDs. The IR aim laser is co-aligned with the visible aim laser. The IR aim laser is boresighted to the weapon using the azimuth and elevation boresight adjusters.

3-17. The IR aim laser is a Class 1 device. The laser can reach a minimum of 100 meters in full moonlight conditions. To operate the IR aim laser, turn the mode selector switch to IR and activate the lasers using the toggle switch or remote cable switch. To operate continuously, move the toggle switch to the Steady-On position or engage by pressing remote cable switch pressure pad.

3-18. To discontinue use, move the toggle switch to the OFF position or release pressure from the remote cable switch pressure pad. The IR aim laser also may be operated in combination with the IR illuminator. To operate in this mode, rotate the mode selector switch to the IR/ILL mode and ensure the laser inactivation switch is in the ON position. Activate the lasers using the toggle switch or remote cable switch.

VISIBLE AIM LASER

3-19. The visible aim laser is used to point at objects and targets during daylight or night conditions without the need for NVDs. The visible aim laser is co-aligned with the infrared aim laser. The visible aim laser is boresighted to the weapon using the azimuth and elevation boresight adjusters.

3-20. The visible aim laser is a Class 3A device and can reach out to a maximum of 25 meters in daylight conditions (not in direct sunlight). To operate—

1. Turn the mode selector switch to visible (VIS).
2. Activate the laser using the toggle switch or remote cable switch.

3-21. To operate continuously, move the toggle switch to the STEADY-ON position or engage by pressing the remote cable switch pressure pad. To discontinue use, move the toggle switch to the OFF position or release pressure from the remote cable switch pressure pad.

3-22. The visible aim laser may also be operated in combination with the white light illuminator. To operate in this mode—

1. Rotate the mode selector switch to the VIS/ILL mode and ensure the laser inactivation switch is in the ON position.
2. Activate the lasers using the toggle switch or remote cable switch.

Aiming Devices

WHITE LIGHT ILLUMINATOR

3-23. The white light illuminator is used to illuminate the target area in conditions of darkness without the need for NVDs.

3-24. To operate the white light illuminator turn the mode selector switch to VIS/ILL and activate the lasers using the toggle switch or remote cable switch. To operate continuously, move the toggle switch to the STEADY-ON position or engage by pressing remote cable switch pressure pad. To discontinue use, move the toggle switch to the OFF position or release pressure from the remote cable switch pressure pad.

3-25. In the VIS/ILL mode, the white light illuminator operates in combination with the visible aiming laser. To deactivate the visible aiming laser, and operate in white light illuminator only, rotate the laser inactivation switch to the OFF position. Activate the white light illuminator using the toggle switch or remote cable switch.

3-26. White light illumination focus can be achieved by rotating the bezel reflector assembly (small ring at top of bezel mount assembly) with thumb and forefinger in a clockwise or counterclockwise manner. The bezel reflector assembly can be rotated freely to obtain the desired white light beam size.

PATTERN GENERATORS

3-27. The pattern generators may be used in front of the infrared aiming beams to create unique patterns of projected laser energy. There are five different pattern generators, each contained within a shroud that installs over the front of the AN/PEQ-14 body, and is secured in place using the adjuster tool. Figure 3-5 on page 3-12, illustrates the patterns that are produced: circle, square, and triangle, plus sign, and T shape. Each shape is labeled on the side of the shroud.

Chapter 3

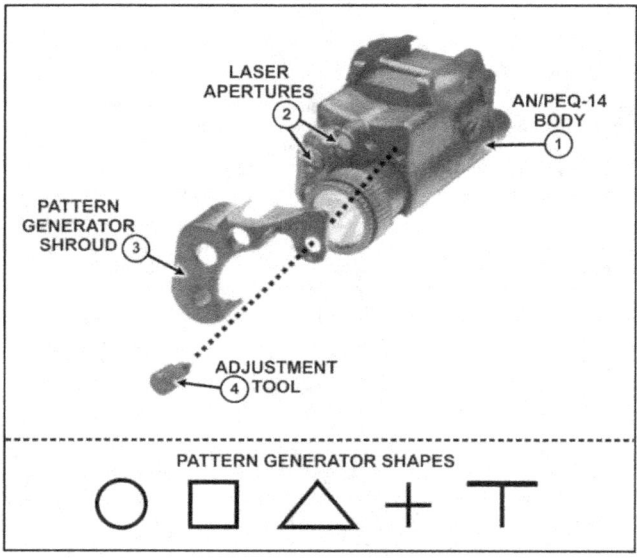

Figure 3-5. Pattern generator install

INSTALLATION OF PATTERN GENERATOR SHROUD

3-28. To install the pattern generator shroud, follow the procedures listed as follows (and see figure 3-5):

- To install a pattern generator shroud, remove the adjuster tool by turning counterclockwise.
- Align the shroud with the AN/PEQ-14 laser apertures and place onto AN/PEQ-14 body.
- Replace the adjuster tool into the AN/PEQ-14 body by threading clockwise to secure shroud in place.
- Ensure the adjuster tool is tight to prevent it from becoming loose or falling out during weapon firing.

Chapter 4
Holsters and Accessories

This chapter provides the principles of operation of a holster and pistol accessories, and provides general information concerning their capabilities, use, and how they function.

REQUIREMENTS

4-1. Units must provide Soldiers with a holster that has a high level of retention, achieved by a friction-based grip; a mechanical retention release such as a thumb-operated lever; and mounting hardware appropriate to the situation. The holster should also protect the pistol from damage and debris. Figure 4-1 shows authorized holsters for use with the Army service pistol.

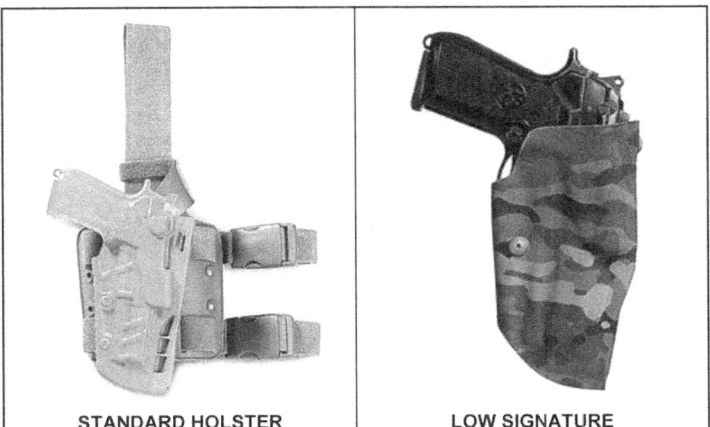

Figure 4-1. Holsters

PLACEMENT

4-2. The proper placement of pistol gear helps ensure safety, and helps a Soldier effectively handle and employ the pistol. A holster helps the Soldier—
- Maintain control of the pistol while conducting tasks that require both hands such as climbing, handling detainees, providing buddy aid, or other tasks.
- Keep the enemy from taking the pistol during close quarters combat.
- Access and manipulate the pistol when necessary, especially when wearing other equipment.
- Conceal the pistol to perform certain jobs.

4-3. The holster should be placed midline on the hip. The Soldier should be able to take their firing arm and extend it to their side and cup their firing hand and ensure it doesn't hit the bottom of the holster. If it does, then the holster is sitting to low and will hinder the draw.

4-4. The Soldier needs to consider the type of holster used and how the placement of the holster can impact the manner in which the pistol is drawn. Lanyards and other tethers can impede use of the pistol. Selection of the correct retention device should address concerns about the pistol becoming loose or falling from the holster.

4-5. The path that the pistol takes when drawing and re holstering MUST be free and clear of all equipment.

4-6. The two most common placements of holsters for the Soldier is on the waistline and thigh. Figure 4-2 on page 4-3 shows waistline and thigh placement examples. Listed below are alternate placement methods that may be necessary due to specific mission requirements. Soldiers must modify their draw and practice from these alternate placement positions:
- Chest (drivers).
- Under shoulder (pilots and tank crew members).
- Direct mount to IOTV (Soldiers wearing body armor).
- Concealment (Soldiers requiring a low signature).

Holsters and Accessories

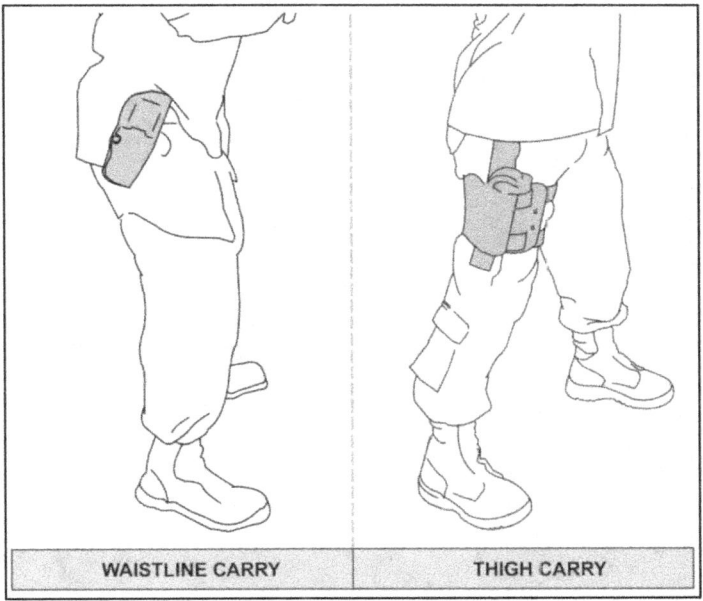

Figure 4-2. Placement

Chapter 4

ACCESSORIES

4-7. The Soldier can mount a light on the pistol. This allows use of the pistol in low light situations. Appendix B provides more information on how to use the weapon-mounted light. (See figure 4-3.) Table 4-1 provides ordering information for supply to purchase lights and adapters. To accommodate a weapon-mounted light, unit supply will need to purchase a holster that will fit over the light and adapter. This may require the purchase of a commercial holster.

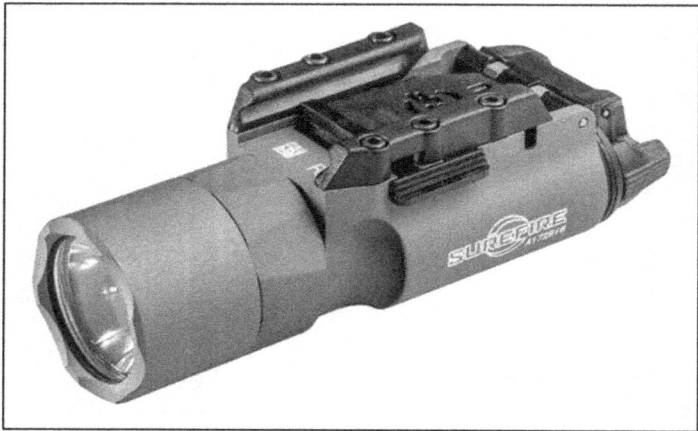

Figure 4-3. Weapon-mounted light

Table 4-1. Ordering information for M9 accessories

Weapon Light X300 Ultra Light Emitting Diode (LED)	NSN 6230-01-617-8332
MR11 Adapter for M9	NSN1005-01-631-4924

Chapter 5
Employment

This chapter discusses employing the shot process, acquiring the target, drawing and presenting the pistol, defeating the safety, and reholstering the pistol.

FIRING SITUATIONS

5-1. Every Soldier must adapt to the firing situation, integrate the rules of firearms safety, manipulate the fire control, and instinctively know when, how, and where to shoot. The Soldier's ability to hit the target under conditions of extreme stress rely upon the following:
- Interpret and act upon perceptual cues related to the target, front and rear sights, pistol movement, and body movement.
- Execute minute movements of the hands, elbows, legs, and feet.
- Coordinate gross motor control of their body positioning with fine motor control of the trigger finger.

5-2. Regardless of the direct fire weapon system, the goal of shooting remains constant: well-aimed shots. To achieve this end state the Soldier's must master sight alignment, sight picture, and trigger control, which are defined below:
- Sight alignment is the relationship between the aiming device and the firer's eye. To achieve proper and effective aim, focus on the front sight post. The Soldier must maintain sight alignment throughout the aiming process.
- Sight picture is the placement of the aligned sights on the target.
- Trigger control is the skillful manipulation of the trigger that causes the pistol to fire without disturbing the aim.

SHOT PROCESS

5-3. The shot process is the basic outline of an individual engagement sequence all firers consider during an engagement, regardless of the weapon employed. The shot process formulates all decisions, calculations, and actions that lead to taking the shot. The shot process may be interrupted at any point before the disengaging and firing of the weapon should the situation change.

5-4. The shot process has three distinct phases:
- Pre-shot.
- Shot.
- Post-shot.

5-5. To achieve consistent, accurate, well-aimed shots, Soldiers must understand and correctly apply the shot process. The sequence of the shot process does not change. However, the application of each element may vary based on the conditions of the engagement.

5-6. Every shot that the Soldier takes has a complete shot process. Grouping, for example, is simply moving through the shot process several times in rapid succession.

5-7. The shot process allows the Soldier to focus on one cognitive task at a time. The Soldier must maintain the ability to mentally organize the tasks and actions of the shot process into a disciplined mental checklist, and focus their attention on activities that produce the desired outcome–a well-aimed shot.

5-8. The level of attention allocated to each element during the shot process is relative to the conditions of each individual shot. Table 5-1 provides an example of a shot process.

Table 5-1. Shot process example

	POSITION
PRE-SHOT	Natural Point of Aim
	Sight Alignment/Picture
	Hold
SHOT	Refine Aim
	Trigger Control
POST-SHOT	Follow-through
	Recoil management
	Call the Shot
	Evaluate

5-9. Functional elements of the shot process are the linkage between the soldier, the weapon, the environment, and the target that directly impact the shot process and ultimately the consistency, accuracy, and precision of the shot. When used appropriately, they build a greater understanding of any engagement.

5-10. The functional elements are interdependent. An accurate shot, regardless of the weapon, requires the soldier to establish, maintain, and sustain the following elements:
- *Stability.* The soldier stabilizes the weapon to provide a consistent base to fire from and maintain through the shot process until the recoil pulse has ceased. This process includes how the soldier holds the weapon, uses structures or objects to provide stability, and the soldier's posture on the ground during an engagement.
- *Aim.* Aim is the continuous process of orienting the weapon correctly, aligning the sights, aligning on the target, and the appropriate lead and elevation (hold) during a target engagement.

Employment

- *Control.* Control entails all the conscious actions of the soldier before, during, and after the shot process that the soldier specifically is in control of. The first of which is trigger control. This includes whether, when, and how to engage. It incorporates the soldier as a function of safety, as well as the ultimate responsibility of firing the weapon.
- *Movement.* Movement is the process of the soldier moving during the engagement process. It includes the soldier's ability to move laterally, forward, diagonally, and in a retrograde manner while maintaining stabilization, appropriate aim, and control of the weapon.

5-10. The elements of the shot process define the tactical engagement that require the soldier to make adjustments to determine appropriate actions, and compensate for external influences on their shot process. When all elements are applied to the fullest extent, soldiers will be able to rapidly engage targets with the highest level of precision.

5-11. The most complex form of engagement is under combat conditions when the soldier is moving, the enemy is moving, under limited visibility conditions, or a combination of the three. Soldiers and leaders must continue to refine skills and move training from the simplest shot to the most complex. Applying the functional elements during the shot process builds a firer's speed while maintaining consistency, accuracy, and precision during complex engagements.

TARGET ACQUISITION

5-12. Target acquisition is the ability of a Soldier to rapidly recognize threats to the friendly unit or formation. It is a critical Soldier function before any shot process begins. It includes the Soldier's ability to use all available optics, sensors, and information to detect potential threats as quickly as possible.

5-13. Target acquisition requires the Soldier to apply an acute attention to detail in a continuous process based on the tactical situation. The target acquisition process includes all the actions a Soldier must execute rapidly, which are—
- *Detect* potential threats (target detection).
- *Identify* the threat as friend, foe, or noncombatant (target identification).
- *Prioritize* the threat(s) based on the level of danger they present (target prioritization).

5-14. Soldiers must master a series of skills to perform effective target detection. Detection is an active process during combat operations with or without a clear or known threat presence. The Soldier's detection skills enable all engagements and are built upon the following three skill sets:
- *Scan and search.* Scan and search is a rapid sequence of various techniques to identify potential threats. Soldier scanning skills determine potential areas where threats are most likely to appear.
- *Acquire.* Acquire is a refinement of the initial scan and search, based on irregularities in the environment.

- *Locate.* Locate is the Soldier's ability to determine the general location of a threat and to engage with accuracy or to inform the small-unit leader of contact with a potential threat.

TARGET IDENTIFICATION

5-15. Identifying (or discriminating) a target as friend, foe, or noncombatant (neutral) is the second step in the target acquisition process. The identification process is complicated by the increasing likelihood of having to discriminate between friend/foe and combatant/noncombatant in urban settings or restricted terrain. To mitigate fratricide and unnecessary collateral damage, Soldiers use all of the situational understanding tools available and develop tactics, techniques, and procedures for performing target discrimination.

5-16. The Soldier must be able to positively identify the threat as one of the following three classifications:
- *Friend.* Any force, U.S. or allied, that is jointly engaged in combat operations against an enemy of the U.S. in a theater of operation.
- *Foe (enemy combatant).* Any individual who has engaged acts against the U.S. or its coalition partners in violation of the laws and customs of war during an armed conflict.
- *Noncombatants.* Personnel, organizations, or agencies that are not taking a direct part in hostilities. Noncombatants include individuals such as medical personnel, chaplains, United Nations observers, media representatives, or those out of combat such as the wounded or sick. The Red Cross or Red Crescent are examples of organizations classified as noncombatants.

TARGET PRIORITIZATION

5-17. The Soldier must prioritize each target and carefully plan the burst to ensure successful target engagement when faced with multiple targets. The keys to a successful engagement of multiple targets are the Soldier's mental preparedness and the ability to make split-second decisions. The proper mindset allows the Soldier to react instinctively and control the pace of the battle rather than reacting to the adversary threat.

Employment

5-18. Targets are prioritized into three threat levels—
- *Most dangerous*. A threat that has the capability to defeat the friendly force and is preparing to do so. These targets must be defeated immediately.
- *Dangerous*. A threat that has the capability to defeat the friendly force, but is not prepared to do so. These targets are defeated after all most dangerous targets are eliminated.
- *Least dangerous*. Any threat that cannot defeat the friendly force, but can coordinate with other threats that are more prepared. Least dangerous targets are defeated after all threats of a higher threat level are defeated.

DRAW AND PRESENT THE PISTOL

5-19. An efficient draw allows crucial seconds to refine aim. The draw is the manipulation by which a shooter removes a pistol from its holster, and presentation is the manipulation by which a shooter drives the pistol to the target. The transition between these two actions is fluid and seamless; their purpose involves removing the pistol from the holster and bringing it to bear on the target as quickly as possible. Figures 5-1a through 5-1e (pages 5-6 through 5-10) illustrate the five steps of the draw and presentation.

Note. Draw varies based on holster type and Soldier experience.

Chapter 5

STEP 1: PREPARE TO DRAW
    a. Move hand to pistol.
    b. Establish high, firm grip.
    c. Prepare to defeat holster's retention device.
    d. Anchor the nonfiring hand to body in preparation to receive the pistol.
    e. Assume a correct stance.
        (1) Place feet about shoulder's width apart.
        (2) Slightly bend the knees.
        (3) Place weight on balls of the feet.

Figure 5-1a. Prepare to draw

Employment

STEP 2: GRIP AND DEFEAT
a. After acquiring grip, defeat holster's retention device and swiftly extract pistol from holster in a straight upward motion.
b. Keep nonfiring hand anchored to the body in preparation to receive the pistol.

Figure 5-1b. Grip and defeat

Chapter 5

STEP 3: DRAW AND ROTATE

a. As soon as the muzzle of the pistol clears the holster, drop the elbow of your firing hand and rotate the pistol to orient the muzzle toward the target.
b. Move the nonfiring hand into position to support the pistol.

Figure 5-1c. Draw and rotate

Employment

STEP 4: MEET AND GREET
a. Slide the fingers of the nonfiring hand under and against the trigger guard.
b. Rotate the nonfiring hand so that the heel of the nonfiring hand is against the pistol grip, resting between the space provided by the fingers and heel of the firing hand.
c. Ensure the firing hand is high on the pistol grip with the finger off the trigger.
d. Defeat the safety.

Figure 5-1d. Meet and greet

Chapter 5

STEP 5: EXTEND AND PREPARE
a. Bring the pistol up to your sight line (sight line is when your sights and target are aligned).
b. Ensure the thumb of your firing hand is on top of your other (nonfiring) thumb.
c. Lock out your nonfiring hand's wrist.
d. As you are pushing the pistol out, prepare the trigger.
e. At full presentation, ensure you have the following:
   (1) A high, firm grip.
   (2) Correct sight alignment.
   (3) Correct sight picture.
   (4) Proper stance.

Figure 5-1e. Extend and prepare

Employment

DISENGAGE THE SAFETY

5-20. A Soldier can disengage the safety at various points during draw and presentation. Figure 5-2 shows methods for defeating the safety. The figure below is in no specific order.

5-21. An important feature of the M9 service pistol is that it is safe to carry the pistol with the safety/decocking lever in the up (off) position. The M9 service pistol is designed so it will not fire unless the trigger is pulled. Local leadership will make the decision for their Soldiers about the position of the safety/decocking lever while the pistol is in the holster.

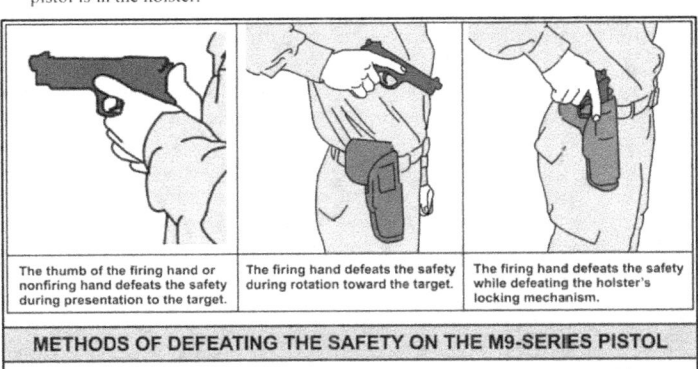

| The thumb of the firing hand or nonfiring hand defeats the safety during presentation to the target. | The firing hand defeats the safety during rotation toward the target. | The firing hand defeats the safety while defeating the holster's locking mechanism. |

METHODS OF DEFEATING THE SAFETY ON THE M9-SERIES PISTOL

A shooter can defeat the safety at various points during draw and presentation. However, the safety should be defeated before the arms extend away from the body.

Figure 5-2. Methods for defeating the safety

REHOLSTER THE PISTOL

5-22. Once firing is completed, Soldiers should reholster their pistols. Before reholstering, Soldiers must decock their pistols using the decocking lever. This mechanism safely returns the pistol to a safe condition. When reholstering, the Soldier should ensure his or her finger is off the trigger and outside of the trigger guard. The Soldier should not attempt to lower the hammer using his or her thumb. There is no time limit for reholstering; this should be done without rushing and safely while observing the pistol and holster.

This page intentionally left blank.

Chapter 6
Stability

This chapter covers how the Soldier carries and grips the pistol; and uses structures and the Soldier's own body to create a stable firing position. Stability is provided through four functions: support, muscle relaxation, natural point of aim, and recoil management. These functions provide the Soldier the means to best stabilize their weapon system during the engagement process.

SUPPORT

6-1. Support can be natural or artificial or a combination of both. Natural support comes from a combination of the Soldier's bones and muscles. Artificial support comes from objects outside the Soldier's body. The more support a particular position provides, the more stable the weapon. For the pistol, the majority of the support will come from the Soldier having a solid stance and grip, as most pistol engagements will be from the standing position.

LEG POSITION

6-2. The position of the legs varies greatly depending on the firing position and Soldier discretion. The position may require the legs to support the weight of the Soldier's body, support the nonfiring elbow, or to meet other requirements for the firing position. For example, standing supported, beyond the act of standing, the Soldier's rear leg provides pressure forward. When standing unsupported, the body is upright with the legs staggered and knees slightly bent. In the prone, the firer's legs may be spread apart flat on the ground or bent at the knee.

STANCE OR CENTER OF GRAVITY

6-3. The physical position of a Soldier before, during, and after the shot that relates to the firer's balance and posture. The stance or center of gravity does not apply when firing from the prone position.

6-4. The stance or center of gravity specifically relates to the Soldier's ability to maintain the stable firing platform during firing, absorbing the recoil impulses, and the ability to aggressively lean toward the target area during the shot process.

GRIP

6-5. A proper grip provides the Soldier maximum control of the pistol. The pistol must become an extension of the hand and arm; it should replace the finger in pointing at an object.

Chapter 6

FIRING HAND

6-6. Proper placement of the firing hand improves trigger control. Place the pistol grip in the "V" formed between the thumb and index finger. The pressure applied is similar to a firm handshake grip. Different Soldiers have different size hands and lengths of fingers, so there is no set position of the finger on the trigger, but Soldiers should try to place the trigger finger on the trigger between the tip and second joint so that it can be squeezed to the rear. The trigger finger must work independently on the remaining fingers.

NONFIRING HAND

6-7. Proper placement of the nonfiring hand is based on the firing position and placement of the nonfiring elbow to provide the stability of the weapon. Placement is adjusted during supported and unsupported firing to maximize stability.

MUSCLE RELAXATION

6-8. Muscle relaxation is the ability of the Soldier to maintain orientation of the weapon appropriately during the shot process while keeping the major muscle groups from straining to maintain the weapon system's position. Relaxed muscles contribute to stability provided by support:
- Strained or fatigued muscles detract from stability.
- As a rule, the more support from the Soldier's bones, the less is required from the muscles.
- The greater the skeletal support, the more stable the position, because bones do not fatigue or strain.
- As a rule, the less muscle support required, the longer the Soldier can stay in position.

NATURAL POINT OF AIM

6-9. The natural point of aim is the point where the barrel naturally orients when the Soldier's muscles are relaxed and support is achieved. The natural point of aim is built upon the following principles:
- The closer the natural point of aim is to the target, the less muscle support is required.
- The more stable the position, the more resistant to recoil it is.
- More of the Soldier's body on the ground equals a more stable position.
- More of the Soldier's body on the ground equals less mobility for the Soldier.

6-10. When a Soldier aims at a target, the lack of stability creates a wobble area, where the sights oscillate slightly around and through the point of aim. If the wobble area is larger than the target, the Soldier requires a steadier position or a refinement to their position to decrease the size of his or her wobble area before trigger squeeze.

Note. The steadier the position, the smaller the wobble area. The smaller the wobble area, the more precise the shot.

RECOIL MANAGEMENT

6-11. Recoil management is the result of a Soldier assuming and maintaining a stable firing position which mitigates the disturbance of one's sight picture during the cycle of function of the weapon.

6-12. The Soldier's firing position manages recoil using support of the weapon system, the weight of their body, and the placement of the weapon during the shot process. Proper recoil management allows the sights to rapidly return to the target and allows for faster follow up shots.

ASPECTS OF GRIP

6-13. When practicing to achieve a proper grip, the Soldier should consider the following aspects of a good grip:

RECOIL

6-14. Recoil causes the pistol to rise or flip beyond the Soldier's control, causing the Soldier to wait before the sights align with the target again. Gripping the pistol properly enables the Soldier to have the fullest control over recoil and helps drive the pistol to any subsequent targets. The Soldier must control the force of recoil by transmitting it straight to the rear into both the nonfiring arm and firing arm.

6-15. Soldiers need to focus on the tip of the front sight when the pistol recoils to aid in acquiring sights faster for follow up shots.

PRESSURE

6-16. Applying the proper amount and direction of pressure helps mitigate errors as the pistol is locked into position. A Soldier's grip should provide counteracting planes of pressure to keep the pistol stable throughout trigger manipulation and recoil. Figures 6-1a and 6-1b on page 6-4 show the multiple planes of pressure used in one-handed and two-handed grips.

Chapter 6

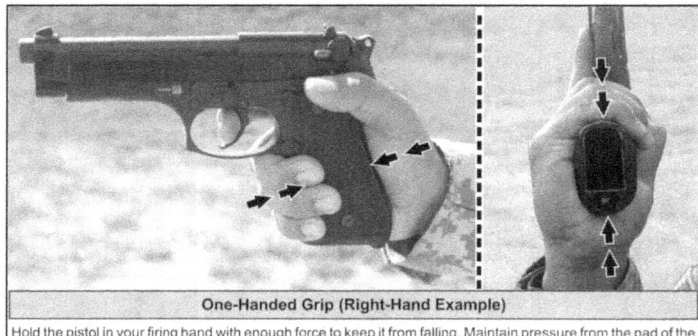

Hold the pistol in your firing hand with enough force to keep it from falling. Maintain pressure from the pad of the fingers (the force) to the heel of the hand (counterforce).

Figure 6-1a. Force and counterforce in the pistol grip (one-hand grip)

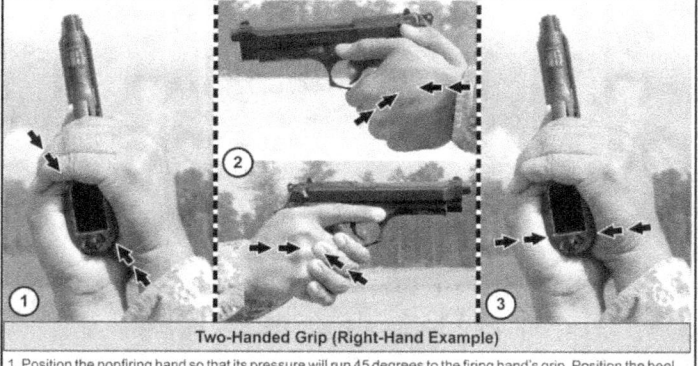

1. Position the nonfiring hand so that its pressure will run 45 degrees to the firing hand's grip. Position the heel against the pistol grip and the fingers over the firing hand. The fingers will provide the force to the heel's counterforce.

2. Extend the thumb and roll the wrist forward, putting downward pressure on the muzzle of the pistol (force). Use the heel of the hand to put pressure on the back strap of the grip (counterforce). This movement assists in recoil management.

3. Press the heels of both hands together as the arms are extended during presentation of the pistol. Each heel will exert force/counterforce against the other.

Figure 6-1b. Force and counterforce in the pistol grip (two-hand grip)

Stability

FIRMNESS

6-17. Soldiers should grip the pistol firmly enough while firing a shot so that the pistol does not shift or slip in their hands, but not so firmly that the muscles of the hand and forearm begin to fatigue. Soldiers should use only that force required to support the pistol and hold it firmly in the firing hand; the firmness of the firing hand simply keeps the pistol from falling to the ground.

COMFORT

6-18. The grip must be as comfortable as possible. After the hand becomes accustomed to the added stress, the muscles of the hand and lower arm should experience no discomfort from the way the pistol is placed in the hand. If the grip is awkward the hand muscles will fatigue.

RELAXATION

6-19. Holding the grip too long without occasional relaxation results in early fatigue, and fatigue disrupts control.

CONSISTENCY

6-20. Ideally, a Soldier tries to maintain a consistent grip. In terms of grip, consistency means that no pressures are added to the pistol during firing and the Soldier maintains his or her hand placement throughout shooting.

6-21. Inconsistencies in the grip will produce different results over all aspects of the shot from sight alignment to recoil management.

UNIFORMITY

6-22. Soldiers should grip the pistol uniformly and in the same manner each time. The grip should not vary from one shot to the next, from one series of shots to the next, or from one day's shooting to the next. The tightness of the grip should not change, as varying the grip pressure adversely affects sight alignment and sight picture. Any tightening or loosening of the grip from an established one can cause the sights to move out of alignment.

Chapter 6

LEVERAGE

6-23. Proper leverage helps a Soldier manage recoil and initiate faster follow-on shots. To manage this force, the Soldier should grip the pistol high on the grip as close as possible to the line of bore. (See figure 6-2.) The thumb of the nonfiring hand should parallel the pistol frame. This lets the wrist of the nonfiring side "lock out" and keeps the pivot point of the applied force close to and high in relation to the barrel.

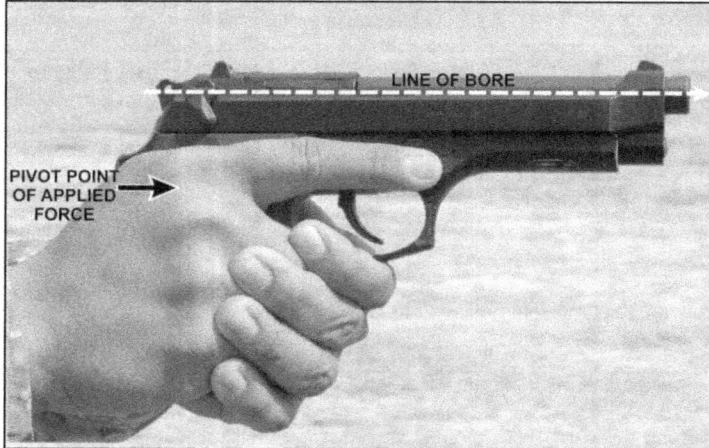

Figure 6-2. Pistol leverage

Stability

ONE- AND TWO-HAND GRIPS

6-24. A proper grip provides the Soldier maximum control of the pistol. A Soldier can grip a pistol using a one or two handed grip.

ONE-HANDED GRIP

6-25. The one-handed grip can be used upon the first contact made with the pistol during the draw-stroke or from the pistol transition. Figure 6-3 shows the one-handed grip. A Soldier can use this grip to engage a target in extreme situations, such as when the other hand is injured and the Soldier must continue to engage targets. This grip can serve as a training tool for learning correct trigger control.

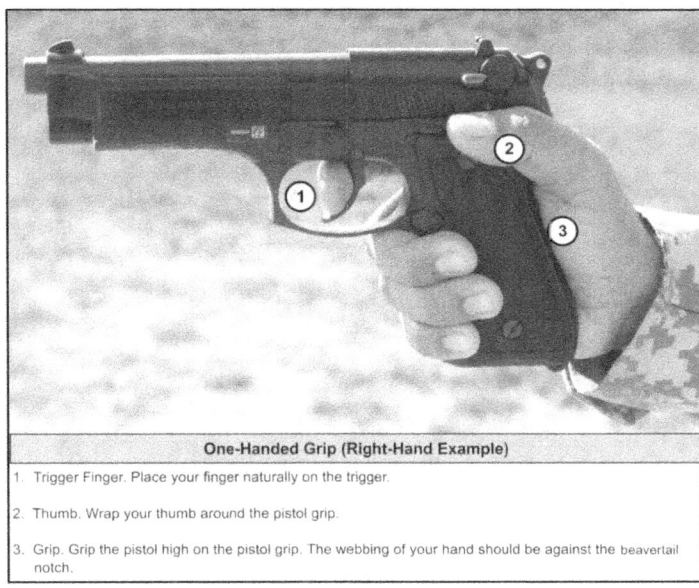

One-Handed Grip (Right-Hand Example)
1. Trigger Finger. Place your finger naturally on the trigger.
2. Thumb. Wrap your thumb around the pistol grip.
3. Grip. Grip the pistol high on the pistol grip. The webbing of your hand should be against the beavertail notch.

Figure 6-3. One-handed grip (right-hand example)

TWO-HANDED GRIP

6-26. The two-handed grip allows the Soldier to steady his or her firing hand and provide maximum support while firing. (See figure 6-4.)

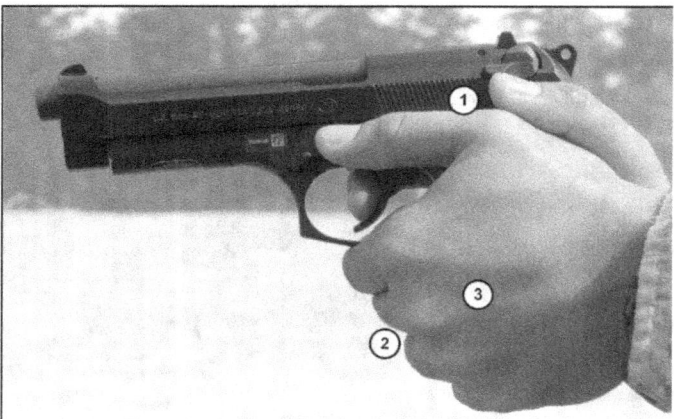

Two-Handed Grip (Right-Hand Example)

1. Thumbs. Place the nonfiring thumb alongside the firing thumb. The firing thumb should sit above the nonfiring thumb. Both thumbs sit parallel to the pistol frame to avoid interfering with the pistol's slide. The thumbs of both hands will generally appear to be stacked on top of each other.

2. Fingers of the nonfiring hand. Firmly close the fingers of nonfiring hand over the fingers of firing hand. Ensure the nonfiring index finger is between the middle finger of the firing hand and the trigger guard.

3. Nonfiring hand. Support hand provides most grip pressure. Grip with nonfiring hand high on the pistol grip, without contacting the slide. Ensure palm of nonfiring hand has full contact with the pistol grip an links perfectly with firing hand.

Figure 6-4. Two-handed grip (right-hand example)

Stability

SHOOTER-GUN ANGLE

6-27. The shooter-gun angle is the relationship between the Soldiers body position and the direction of the weapons muzzle. The shooter-gun angle is important to the Soldier; a better body position will aid the Soldier in better recoil management and control of the pistol.

FIELD OF VIEW

6-28. The field of view is the extent that the human eye can see at any given moment. The field of view is based on the Soldier's view. The field of view is what the Soldier sees, and includes the areas where the Soldier can detect potential threats.

CARRY POSITIONS

6-29. The pistol has two carry positions (see figures 6-1a and 6-1b on page 6-4), holstered and ready. These positions may be directed by the leader, or assumed by the Soldier based on holstered

6-30. Soldiers use the holster carry when they need their hands for other tasks and no threat is present or likely. The weapon is holstered and the safety is engaged.

6-31. In the ready position, hold the pistol with the two-handed grip. Hold the upper arms close to the body and the forearms at about a 45-degree angle. Point the pistol toward the target's center while moving forward. Figure 6-5 shows how to assume the carry positions.

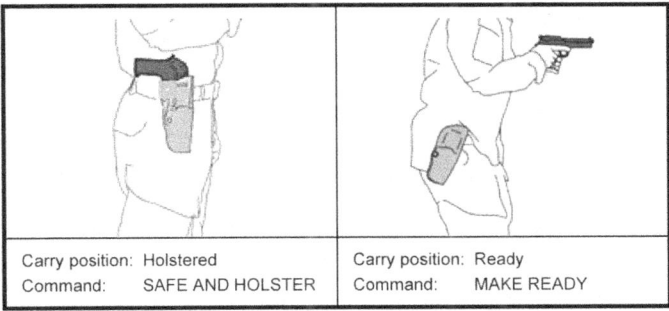

Figure 6-5. Carry positions

Chapter 6

STABILIZATION

6-32. As a rule, positions that are lower to the ground provide a higher level of stability. When the position elevates the level of stability decreases as shown in figure 6-6. Each firing position figure will list the stability rating using a high, moderate or low rating.

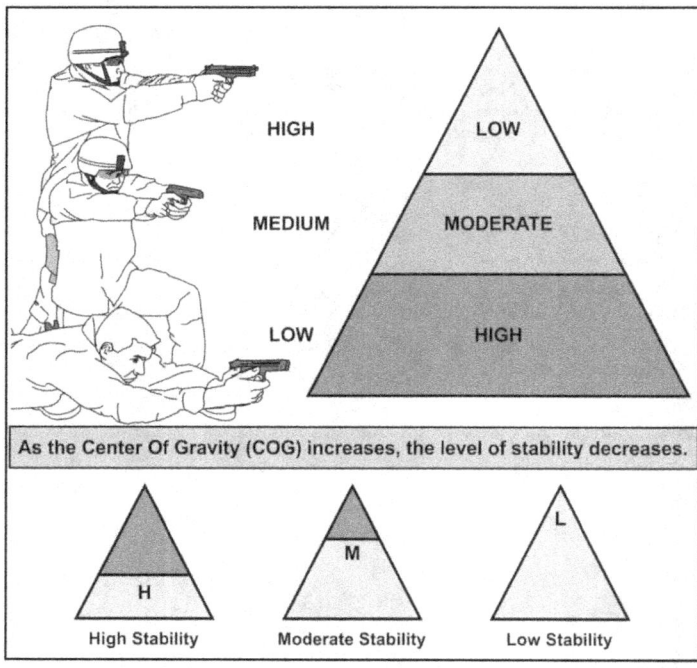

Figure 6-6. Firing stability

Stability

FIRING POSITIONS

6-33. Any position the Soldier assumes should support sight alignment and trigger control, and it should not interfere with the operation of the pistol. The six firing positions are shown from low stability to high stability. The triangle in the bottom right of the firing position figures will list the stability rating.

6-34. Soldiers must not completely lock out their arms when firing the pistol. Having a slight bend in their elbows will aid in managing recoil and control of the service pistol. Soldiers will lock out their wrists to assist in recoil management.

STANDING UNSUPPORTED

6-35. Face the target. (See figure 6-7.) Place feet a comfortable distance apart, about shoulder width. Achieve a two-handed grip with the wrists locked out, and fully extend the arms, without locking out the elbows, toward the targets center. Keep the body straight with the shoulders slightly forward of the buttocks.

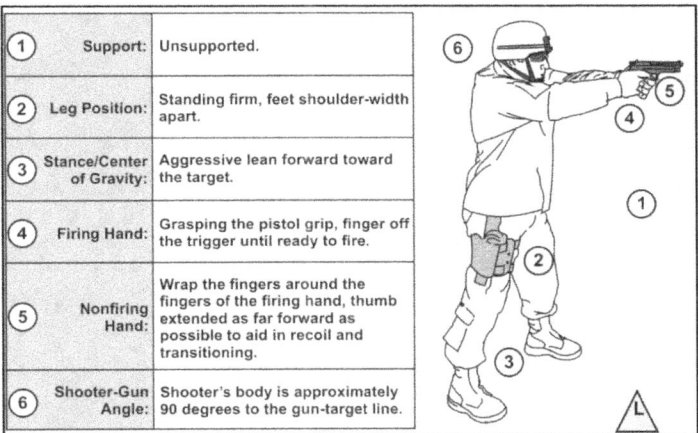

1	Support:	Unsupported.
2	Leg Position:	Standing firm, feet shoulder-width apart.
3	Stance/Center of Gravity:	Aggressive lean forward toward the target.
4	Firing Hand:	Grasping the pistol grip, finger off the trigger until ready to fire.
5	Nonfiring Hand:	Wrap the fingers around the fingers of the firing hand, thumb extended as far forward as possible to aid in recoil and transitioning.
6	Shooter-Gun Angle:	Shooter's body is approximately 90 degrees to the gun-target line.

Figure 6-7. Standing unsupported position

STANDING SUPPORTED POSITION

6-36. Use available cover for support—for example, a wall—or a barricade to stand behind. (See figure 6-8.) Stand behind a barricade with the firing side on line with the edge of the barricade. Place the knuckles of the nonfiring fist at eye level against the edge of the barricade. Lock the wrist, and fully extend the elbow of the firing arm. Align the outside foot with the edge of the barricade.

①	Support:	Supported using available structure.
②	Leg Position:	Standing firm, feet shoulder-width apart. Move the foot of the non-firing side as forward as possible.
③	Stance/Center of Gravity:	Aggressive lean forward toward the target.
④	Firing Hand:	Grasping the pistol grip, finger off the trigger until ready to fire.
⑤	Nonfiring Hand:	Place the knuckles of the non-firing fist at eye level against the edge of the barricade.
⑥	Shooter-Gun Angle:	Stand behind the firing side in line with the edge of the barricade.

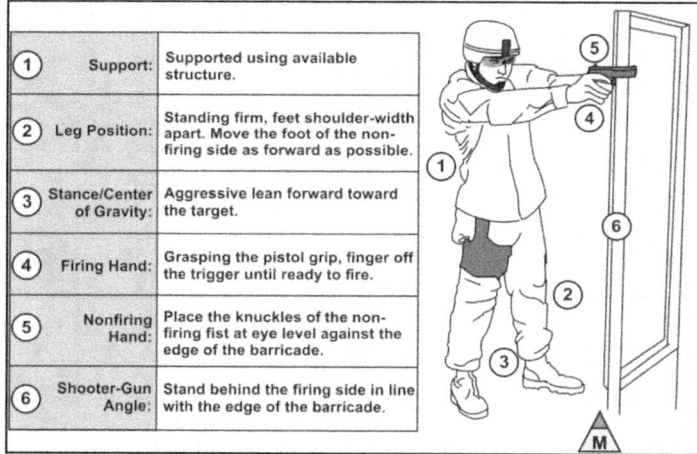

Figure 6-8. Standing supported position

Stability

KNEELING UNSUPPORTED

6-37. In the kneeling unsupported position, ground only your firing-side knee as the main support. (See figure 6-9) Use the two-handed grip for firing. Extend the arms without locking out the elbows. Lock wrists to ensure control and aid in recoil.

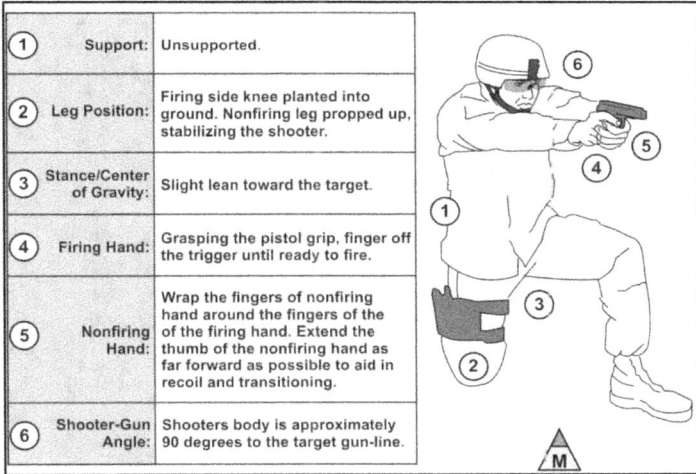

(1)	Support:	Unsupported.
(2)	Leg Position:	Firing side knee planted into ground. Nonfiring leg propped up, stabilizing the shooter.
(3)	Stance/Center of Gravity:	Slight lean toward the target.
(4)	Firing Hand:	Grasping the pistol grip, finger off the trigger until ready to fire.
(5)	Nonfiring Hand:	Wrap the fingers of nonfiring hand around the fingers of the firing hand. Extend the thumb of the nonfiring hand as far forward as possible to aid in recoil and transitioning.
(6)	Shooter-Gun Angle:	Shooters body is approximately 90 degrees to the target gun-line.

Figure 6-9. Kneeling unsupported position

KNEELING SUPPORTED POSITION

6-38. In the kneeling supported position, ground your firing side knee. (See figure 6-10.) Rest your nonfiring arm on your nonfiring knee or seek cover, ground your nonfiring arm on a barricade or vehicle for support. Use the two-handed grip for firing. Extend the firing arm without locking out the elbow and lock the wrist to ensure control.

①	Support:	Supported using available structure or nonfiring knee.
②	Leg Position:	Firing knee on the ground, foot under seat. Nonfiring leg bent approximately 90 degrees outward.
③	Stance / Center of Gravity:	Slight lean toward target.
④	Firing Hand:	Grasping pistol grip, finger off the trigger until ready to fire.
⑤	Nonfiring Hand:	Wrap the fingers of nonfiring hand around the fingers of the firing hand. Extend the thumb of the nonfiring hand as far forward as possible to aid in recoil and transitioning.
⑥	Shooter-Gun Angle:	Shooter body is approximately 90 degrees to the gun-target line.

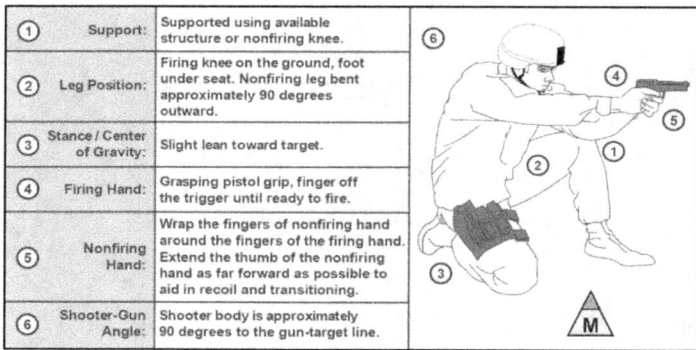

Figure 6-10. Kneeling supported

Stability

PRONE UNSUPPORTED

6-39. Draw, post, and sprawl onto the ground, facing the target. (See figure 6-11.) Extend the arms as much as possible with wrists locked. (Arms may have to be slightly unlocked for firing at high targets.) Wrap the fingers of the nonfiring hand around the fingers of the firing hand. Keep your head down between your arms and behind the weapon as much as possible.

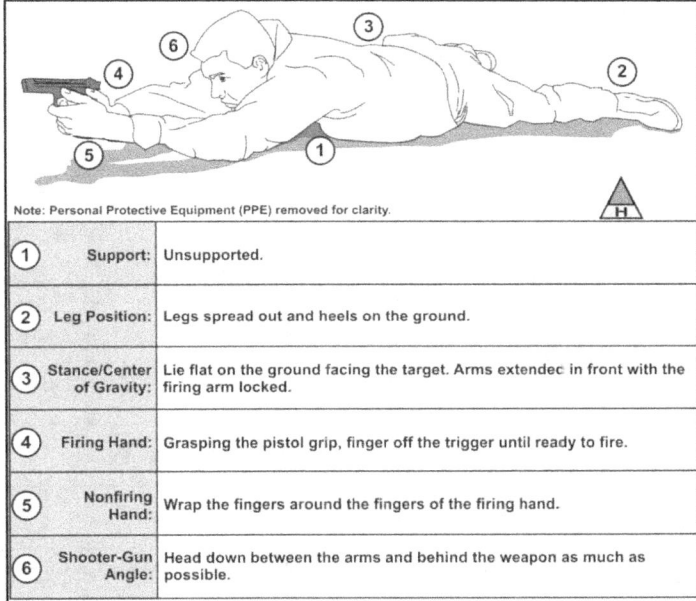

①	Support:	Unsupported.
②	Leg Position:	Legs spread out and heels on the ground.
③	Stance/Center of Gravity:	Lie flat on the ground facing the target. Arms extended in front with the firing arm locked.
④	Firing Hand:	Grasping the pistol grip, finger off the trigger until ready to fire.
⑤	Nonfiring Hand:	Wrap the fingers around the fingers of the firing hand.
⑥	Shooter-Gun Angle:	Head down between the arms and behind the weapon as much as possible.

Figure 6-11. Prone unsupported position

Chapter 6

PRONE SUPPORTED

6-40. Draw, post, and sprawl onto the ground facing the target. (See figure 6-12.) Fully extend the arms in front with the wrists locked out (arms may have to be slightly unlocked for firing at high targets.) Rest the butt of the weapon on the ground for support (or an alternate form of support, such as an assault pack). Wrap the fingers of the nonfiring hand around the fingers of the firing hand. Keep your head down between your arms and behind the weapon as much as possible. Rest cheek on shoulder.

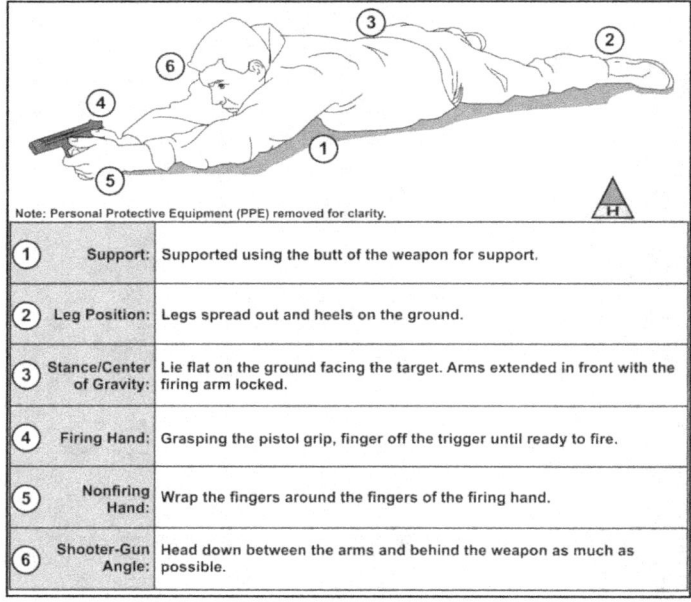

1	Support:	Supported using the butt of the weapon for support.
2	Leg Position:	Legs spread out and heels on the ground.
3	Stance/Center of Gravity:	Lie flat on the ground facing the target. Arms extended in front with the firing arm locked.
4	Firing Hand:	Grasping the pistol grip, finger off the trigger until ready to fire.
5	Nonfiring Hand:	Wrap the fingers around the fingers of the firing hand.
6	Shooter-Gun Angle:	Head down between the arms and behind the weapon as much as possible.

Figure 6-12. Prone supported position

Chapter 7
Aim

The aim element of employment is the continuous process of orienting the weapon correctly, aligning the sights, aligning sights on the target, and applying the appropriate hold during a target engagement. Aiming is a continuous process conducted through pre-shot, shot, and post-shot, to effectively apply lethal fires in a responsible manner with accuracy and precision.

ELEMENTS OF ACCURACY

7-1. During combat, a Soldier balances the need for accuracy with the need for speed. Consider how fast a person can move: If a shooter's draw takes 1 second, someone positioned 20 feet away can reach the shooter before the shooter can fire. Due to the urgency caused by close proximity to the enemy, Soldiers must act fast; they must engage the enemy before the enemy engages them.

7-2. Suppressive fire is any influence (rounds, noise, white light, lasers, and so on) that is placed on the enemy that disrupts their decision making process and alters their ability to place accurate fire on you. Soldiers need to know the difference between precise accuracy, and acceptable accuracy while under stress. The three elements that influence the balance between speed and accuracy are distance to the target, size of the target, and the shooter's ability.

DISTANCE TO TARGET

7-3. The shooter's distance to the target impacts how fast the shooter will engage the target and whether or not the shooter can accept errors in sight alignment. The distance between the target and the shooter is directly related to the refinement of the sight picture.

7-4. During combat, targets can present themselves at various distances. To accurately hit targets as they appear, a shooter requires an understanding of this distance and its impact on their shooting and requires the efficiency of motion and an understanding of the degree of error necessary to increase their tempo when close targets present themselves. Further, the Soldier requires the mental discipline to slow down and refine their shot process when distant targets present themselves.

SIZE OF THE TARGET

7-5. The surface area of the target is directly related to the refinement of the shot process necessary to engage the target. A shooter must have the mental discipline to discern when and how much to compromise the execution of the shot process.

7-6. If the target has a large surface area and is located relatively close the shooter, errors in sight alignment are easier to accept.

SHOOTER'S ABILITY

7-7. A shooter's skill level is the most influential element in determining the balance between speed and accuracy. Shooters should attempt to shoot within their abilities.

COMMON ENGAGEMENTS

7-8. The aiming process for engaging targets consists of the following Soldier actions:
- Weapon orientation. The direction of the weapon as it is held in a stabilized manner.
- Sight alignment. The physical alignment of the sights with the shooters eye.
- Sight picture. The target as viewed through the line of sight.
- Point of aim. The specific location where the line of sight intersects the target.
- Desired point of impact. The desired location of the strike of the round to achieve the desired outcome (incapacitation or lethal strike).

7-9. The aim of the weapon is typically applied to the largest area of any target presented. Sights can be placed on target by utilizing center of visible mass, appropriate hold-off, or on the anticipated or desired point of impact of the round. Each method of placement is unique and requires a complete understanding of the weapon's sights, ammunition, ballistics, target posture, and environmental conditions. The sight pictures used during the shot process are—
- **Pre-shot sight picture.** Encompasses the original point of aim, sight picture, and any holds for the target.
- **Post-shot sight picture.** The Soldier must use this as the point of reference for any sight adjustments for any subsequent shot.

WEAPON ORIENTATION

7-10. The Soldier orients the weapon in the direction of the detected threat. Weapon orientation includes both the horizontal plane (azimuth) and the vertical (elevation) plane. Weapon orientation is complete once the sight and threat are in the Soldier's field of view.

Aim

Horizontal Weapon Orientation

7-11. Horizontal weapon orientation (figure 7-1) covers the frontal arc of the Soldier, spanning the area from the left shoulder, across the Soldier's front, to the area across the right shoulder.

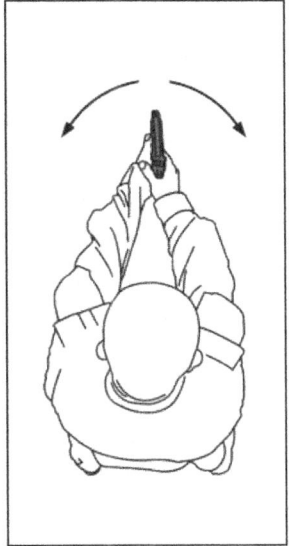

Figure 7-1. Horizontal weapon orientation, example

7-12. Hold your head upright and in a natural position to provide balance and stability. Hold your head level (not tilted to the left or right or forward) to ensure balance so that you can see the target directly in line with the arm and sights. Bring the weapon sights to your head so that your head moves very little or not at all.

7-13. Extend your arms without pushing them past their natural range of motion. Do not lock your elbows. Center your arms on your chest or hold them slightly offset on your dominant eye side, forming a triangle between your chest and arms, if viewed from above.

Chapter 7

Vertical Weapon Orientation

7-14. Vertical weapon orientation (figure 7-2) includes all the aspects of orienting the weapon at a potential or confirmed threat in elevation. This is most commonly applied in restricted, mountainous, or urban terrain where threats present themselves in elevated or depressed firing positions.

Figure 7-2. Vertical weapon orientation, example

7-15. Slightly bend at the waist, with upper torso and shoulders forward of your buckle.

7-16. Set your feet shoulder-width apart, with your nonfiring foot slightly in front of your firing foot. This enables your feet to provide a stable platform to support shooting and recoil management. Distribute your weight between both feet. Focus your weight on the balls of your feet.

Aim

SIGHT ALIGNMENT

7-17. Sight alignment (figure 7-3) is the relationship between the sighting system and the firer's eye.

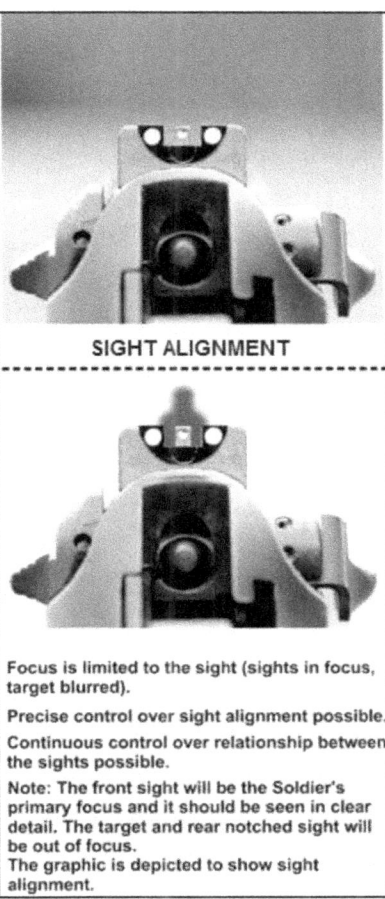

Figure 7-3. Proper sight alignment

SIGHT PICTURE

7-18. The sight picture is the target as viewed through the line of sight, that is, the placement of the aligned sights on the target itself. The Soldier must maintain sight alignment throughout the positioning of the sights.

POINT OF AIM

7-19. The point on the target that the Soldier is using to align their aiming device (iron sights).

7-20. Natural point of aim (NPA) is achieved by adjusting the assumed firing position so that the sights will be aligned with the target with a minimal amount of muscle tension.

DESIRED POINT OF IMPACT

7-21. The point of impact is the location where the projectile strikes the target.

COMMON AIMING ERRORS

7-22. Orienting and aiming a weapon correctly is a practiced skill. Through drills and repetitions, Soldiers build the ability to repeat proper weapons orientation, sight alignment, and sight picture as a function of muscle memory. The most common aiming errors include—

INCORRECT SIGHT ALIGNMENT

7-23. Soldiers may experience this error when failing to focus on the front sight post. Soldiers can also have incorrect sight alignment when using the dots to aim.

INCORRECT SIGHT PICTURE

7-24. This occurs typically when the threat is in a concealed location or moving. This failure directly impacts the Soldier's ability to create and sustain the proper sight picture during the shot process.

NON-DOMINANT EYE USE

7-25. The Soldier gets the greatest amount of visual input from their dominant eye. Eye dominance varies Soldier to Soldier. Some Soldier's dominant eye will be the opposite of the dominant hand. For example, a Soldier who writes with the right hand and learns to shoot pistols right-handed might learn that the dominant eye is the left eye. This is called cross-dominant. Soldiers with strong cross-dominant eyes should consider firing using their dominant eye. Unlike a rifle, it is usually possible to fire a pistol with the opposite eye by simply tilting the head slightly to use the dominant eye. This is preferable to squinting or closing the dominant eye. It is also possible to switch to firing with the non-dominant hand but this may require substantially more training for Soldiers who are not ambidextrous.

INCORRECT ZERO

7-26. Regardless of how well a Soldier aims, if the zero is incorrect, the round will not travel to the desired point of impact without adjustment with subsequent rounds. (AN/PEQ 14 laser can be adjusted). An incorrect zero can also happen with iron sights. The shooter must adjust their hold. Although badly zeroed M9s are not common, it is usually an error in the Soldier's technique, it does happen and the only real solution is for the shooter to hold off. This is often an indication of broken sights.

LIGHT CONDITIONS

7-27. Limited visibility conditions contribute to errors aligning the sight, selecting the correct point of aim, or determining the appropriate hold. Soldiers may offset the effects of low light engagements with image intensifier (I2) devices and the use of laser pointing devices or flashlight. Glare from front sight post can be corrected with sight blackout.

This page intentionally left blank.

Chapter 8
Control

Control is the act of firing the weapon while maintaining proper aim and adequate stabilization until the bullet leaves the muzzle. Trigger control and the Soldier's position work together to allow the sights to stay on the target long enough for the Soldier to fire the weapon and bullet to exit the barrel. This chapter also discusses malfunctions and how to clear malfunctions for the pistol

ARC OF MOVEMENT

8-1. Regardless of how well trained or physically strong a Soldier is, a wobble area (or arc of movement) is present, even when sufficient physical support of the weapon is provided. The arc of movement may be observed as the sights moving in a W shape, vertical (up and down) pulses, circular, or horizontal arcs. The wobble area, or arc of movement, is the extent of lateral horizontal and front-to-back variance in the movement that occurs. (See figure 8-1.)

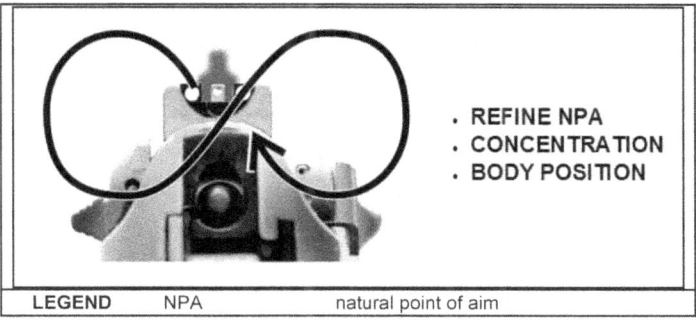

Figure 8-1. Example of arc of movement

Chapter 8

8-2. The control element consists of several supporting Soldier functions, and includes all the actions to minimize the Soldier's induced arc of movement. The Soldier physically maintains positive control of the shot process by managing—
- Trigger control.
- Calling the shot (firing or shot execution).
- Follow-through.

TRIGGER CONTROL

8-3. Trigger control is the act of firing the weapon while maintaining proper aim and adequate stabilization until the bullet leaves the muzzle. Trigger control and the shooter's position work together to allow the sights to stay on the target long enough for the Soldier to fire the weapon and bullet to exit the barrel.

8-4. Stability and trigger control complement each other and are integrated during the shot process. A stable position assists in aiming and reduces unwanted movements during trigger press without inducing unnecessary movement or disturbing the sight picture. A smooth, consistent trigger squeeze, regardless of speed, allows the shot to fire at the Soldier's moment of choosing. When both a solid position and a good trigger squeeze are achieved, any induced shooting errors can be attributed to the aiming process for refinement.
- Trigger finger placement. Correct placement of the trigger finger (see figure 8-2) is necessary to apply pressure on the trigger with the index finger between the tip and second joint (the first bone section). Unnatural trigger finger placement can lead to an inaccurate shot.
- Trigger squeeze. The Soldier pulls the trigger in a smooth consistent manner adding pressure until the weapon fires. Regardless of the speed at which the Soldier is firing the trigger control will always be smooth.
- Trigger reset. Once the Soldier fires a shot, the trigger must be reset prior to firing the next shot. The trigger finger needs to move forward but should maintain contact with the trigger. This will help maintain consistent trigger finger placement. The trigger only needs to move forward far enough to reset. These actions need not be slow and must happen during the recoil of the pistol.

Notes. 1. Do not place the trigger finger on the trigger unless the sights and target are both visible.

2. Do not pin the trigger back; after firing, let the trigger reset.

Control

Figure 8-2. Trigger finger placement

WORKSPACE MANAGEMENT

8-5. The workspace is a spherical area, 12 to 18 inches in diameter centered on the Soldier's chin and about 12 inches in front of his or her chin. The workspace is where the majority of weapons manipulations occur. (See figure 8-3 on page 8-4.)

8-6. Conducting manipulations in the workspace allows the Soldier to keep his or her eyes oriented towards a threat, or the individual sector of fire while conducting critical weapons tasks that require hand-and-eye coordination. Use of the workspace creates efficiency of motion by minimizing the distance the weapon has to move between the firing position to the workspace and return to the firing position.

8-7. Workspace management includes the Soldier's ability to perform the following functions (drills) routinely, with fluidity of motion and smoothness in execution to ensure the fastest possible return of the weapon's orientation to the target area while maintaining observation of the target area or sector of observation.

8-8. The close fight requires rapid manipulations, a balance of speed and accuracy, and very little environmental concerns. Soldiers must move quickly and efficiently through their manipulations of the fire control in order to maintain the maximum amount of muzzle orientation on the threat through the shot process. This second-nature efficiency of movement only comes from regular practice, drills, and repetition.

Chapter 8

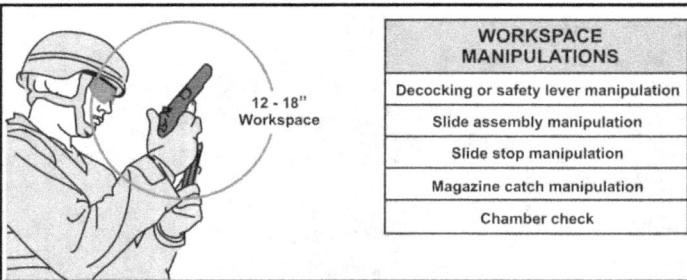

WORKSPACE MANIPULATIONS
Decocking or safety lever manipulation
Slide assembly manipulation
Slide stop manipulation
Magazine catch manipulation
Chamber check

WORKSPACE MANIPULATIONS

- **Decocking or safety lever manipulation.** The ability of the Soldier to change the weapon's status from safe to fire or fire to safe from any position.

- **Slide assembly manipulation.** The ability of the Soldier to smoothly use the slide assembly during operation. This includes any corrective actions to overcome malfunctions, loading, unloading, or clearing procedures.

- **Slide stop manipulation.** The ability of the Soldier to operate the slide stop mechanism on the weapon after shooting.

- **Magazine catch manipulation.** The smooth functioning of the magazine catch during reloading procedures, clearing procedures, or malfunction corrective actions.

- **Chamber check.** The sequence the Soldier uses to verify the status of the weapon's chamber.

Figure 8-3. Workspace management

CALLING THE SHOT

8-9. For accurate shot analysis, the Soldier must know exactly where the sights are when the weapon discharges. Errors such as flinching or jerking of the trigger are visible in the sights before discharge.

8-10. Calling a shot refers to a firer stating exactly where the Soldier thinks a single shot strikes by recalling the sights relationship to the target when the weapon fired. This is normally expressed in clock direction and inches from the desired point of aim.

8-11. The Soldier is responsible for the point of impact of every round fired from their weapon. This requires the Soldier to ensure the target area is clear of friendly and neutral

Control

personnel, in front of and behind the target. Soldiers must also be aware of the environment the target is positioned in, particularly in urban settings—friendly or neutral personnel may be present in other areas of a structure that the projectile can pass through.

FOLLOW-THROUGH

8-12. Follow-through is the sequence of steps required to complete the shot process. Follow-through consists of all actions controlled by the Soldier after the bullet leaves the muzzle:

- Recoil management. This includes the slide assembly recoiling completely and returning to the battery. It includes the unlocking, extraction, ejection, cocking, feeding, chambering and locking of the cycle of function. Any failures indicate a weapon malfunction and require the Soldier to take the appropriate action.
- Recoil recovery. Returning to the position the Soldier was in prior to the shot and reacquiring the sight alignment or sight picture. The Soldier needs to watch the front sight post during recoil.
- Trigger or sear reset. Once the ejection phase of the cycle of function is complete, the weapon initiates and completes the cocking phase. As part of that phase, all mechanical components associated with the trigger, disconnect, and sear are reset. Any failures in the cocking phase indicate a weapon malfunction and require the Soldier to take the appropriate action. The Soldier maintains trigger finger placement and releases pressure on the trigger until the sear is reset, demonstrated by a metallic click. At this point the sear is reset and the trigger pre-staged for a subsequent or supplemental engagement if needed. Don't hold the trigger back on reset. Once the round is fired, let the trigger go forward while maintaining contact with trigger finger.

Note. The longer the trigger is held to the rear, the longer the Soldier prevents the pistol from functioning and delays reengagement.

- Sight alignment/sight picture adjustment. Counteracting the physical changes in the sight alignment or sight picture caused by recoil impulses and returning the sight alignment or sight picture onto the target aiming point.
- Engagement assessment. Once the sight alignment or sight picture returns to the original point of aim, the firer confirms the strike of the round, assesses the target's state, and immediately selects one of the following courses of action:
  - Subsequent engagement. The target requires additional (subsequent) rounds to achieve the desired target effect. The Soldier starts the pre-shot process.
  - Supplemental engagement. The Soldier determines the desired target effect is achieved and another target may require servicing. The Soldier starts the pre-shot process.

- Sector check. All threats have been adequately serviced to the desired effect. The Soldier then checks his or her sector of responsibility for additional threats. If the sector is clear, the Soldier places the weapon on SAFE and applies Rule No. 4. The Soldier continues to scan for threats as the tactical situation requires. The unit's standard operating procedure will dictate any vocal announcements required during the post-shot sequence, for example, the Soldier announcing, "CLEAR," when the sector or room is clear of all threats.
- Correct malfunction. If the firer determines during the follow-through that the weapon malfunctioned during one of the phases of the cycle of function, the Soldier makes the appropriate announcement to the team and immediately executes corrective action.

MALFUNCTIONS

8-13. A malfunction is a break in the cycle of function caused by faulty action of the pistol, ammunition, or Soldier. For the purposes of this book a malfunction is defined as any time the pistol fails to operate as intended. The corrective actions for a malfunction fall into two main categories described below, immediate action and remedial action.

IMMEDIATE ACTION

8-14. Immediate action involves quickly applying a possible correction to a malfunction without determining the actual cause. It does not involve observation, diagnosis of the malfunction, or decision-making beyond recognition that there is a problem. As the term suggests, it is performed immediately and quickly, taking no more than a few seconds. Like a battle drill, it is conducted reflexively, without thought or hesitation. The procedures for performing immediate action in sequence follow:

1. Ensure the decocking lever on the pistol is in the FIRE position.
2. Remove trigger finger from the trigger and ensure it is straight and pressed along the frame.
3. Bring pistol back into workspace.
4. Rotate pistol.
5. With the heel of the nonsupporting hand, forcefully tap upward onto the baseplate of the magazine.
6. Rotate the pistol to observe the chamber and rack the slide.
7. Allow the chambered round to extract and fall to the floor.
8. Observe the chambered round.
9. Rotate pistol back to target and squeeze trigger.

Control

REMEDIAL ACTION

8-15. Remedial action is a conscious, observed attempt to determine the cause of a malfunction and correct it using a specific set of actions. It differs from immediate action in that it requires a Soldier to consciously analyze the status of the weapon to determine the problem and select the appropriate actions to correct it. The procedures for performing remedial action for majority of malfunctions is—

1. Observe the pistol to identify the cause of the malfunction.
2. Keep the pistol pointed at the intended target.
3. Cant the pistol upward to observe the position of the slide.
4. If the slide is locked to the rear, observe the pistol to see if ammunition is present in the magazine.
5. If no ammunition is present, drop magazine and insert a new loaded magazine.
6. If something is obstructing the chamber or keeping the slide from moving fully forward, lock the slide to the rear.
7. Forcefully remove the magazine from the pistol, clear the obstruction, insert a new loaded magazine, send the slide forward, acquire sights on a target, squeeze the trigger, and try to fire again.

8-16. When any weapon fails to complete any phase of the cycle of function correctly, a malfunction has occurred. When a malfunction occurs, the Soldier's priority is to defeat the target as quickly as possible. The Soldier controls which actions must be taken to ensure the target is defeated as quickly as possible based on the threat presented. Soldiers must take into consideration—

- Targets less than 25 meters. Soldier transitions to secondary weapon for the engagement. If no secondary weapon is available, move to a covered and concealed position (if applicable), to correct the malfunction.
- Soldiers quickly move to a covered and concealed position (if applicable), announce their status to their team members, and execute corrective action.

Note. Appendix C covers rifle-to-pistol transitions in greater detail.

This page intentionally left blank.

Chapter 9
Movement

The movement functional element is the process of the Soldier moving tactically during the engagement process. It includes the Soldier's ability to move laterally, forward, diagonally, and in a retrograde manner while maintaining stabilization, appropriate aim, and control of the weapon.

MOVEMENT TECHNIQUES

9-1. Tactical movement of the Soldier is classified in two ways: vertical and horizontal. Each requires specific considerations to maintain and adequately apply the other functional elements during the shot process.

9-2. Soldiers use vertical movements to change firing posture or negotiate terrain or obstacles while actively seeking, orienting on, or engaging threats. Vertical movements are covered in Chapter 6 in discussions about stability. Vertical movements include actions taken to—
- Change between any of the primary firing positions; standing, kneeling, or prone.
- Negotiate stairwells in urban environments.
- Travel across sloping surfaces, obstacles, or terrain.

9-3. Soldiers use horizontal movements to negotiate the battlefield while actively seeking, orienting on, or engaging threats. The Soldier can use one or all of the horizontal movement techniques listed below while keeping the weapon oriented on the threat. (See Figure 9-1 on page 9-2.)
- Forward – movement in a direction directly toward the adversary.
- Retrograde – movement rearward, in a direction away from the threat while maintaining weapon orientation on the threat.
- Lateral left or right – lateral, diagonal, forward, or retrograde movement to the right or left.
- Turning left, right, or about – actions taken by the Soldier to change the weapon orientation left, right, or to the rear. (Soldier reorients weapon first, then self).

Note. If you have to run, stop, and shoot, holster the pistol before taking off. Then, when you begin to slow down, you can redraw and present the pistol, resume a stable position, and reenter a controlled movement.

Chapter 9

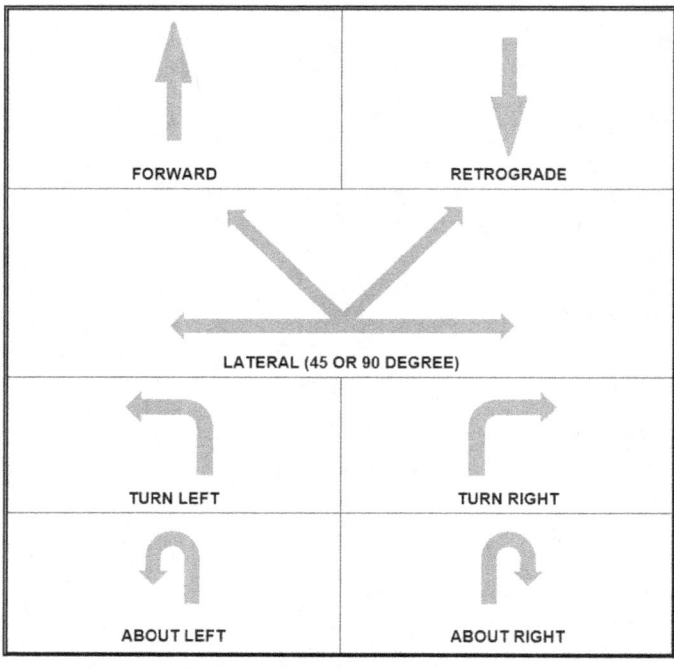

Figure 9-1. Horizontal movements

FORWARD MOVEMENT

9-4. Forward movement is continued progress in a direction toward the threat or route of march. This is the most basic form of movement during an engagement. During forward movement—
1. Roll the foot heel to toe to best provide a stable firing platform.
2. Shoot while maintaining your natural walking gait.
3. Keep the weapon ready.
4. Maintain situational awareness.
5. When moving, bring the weapon back slightly toward your body to aid in recoil and control.

Movement

6. Maintain an aggressive posture.
7. Walk in a straight line, keeping your center of gravity low (drop your hips).
8. Keep the muzzle of the weapon downrange toward the target.
9. Use the upper body as a turret.
10. Twist at the waist.
11. Maintain proper posture.

RETROGRADE MOVEMENT

9-5. Retrograde movement is moving backward while keeping the weapon oriented forward.

Note. Ideally, you would not want to move backwards, but if you have to, these steps should reduce the risk.

9-6. During retrograde movement—
   1. Place your feet toe to heel and drop your center of gravity by bending your knees.
   2. Be careful of tripping when walking backwards.
   3. Ensure all movement is smooth and steady to maintain stability.
   4. When moving, bring the weapon back slightly toward your body to aid in recoil and control.
   5. Keep the muzzle of the weapon downrange toward the target.
   6. Use the upper body as a turret.
   7. Twist at the waist.
   8. Maintain proper posture.

LATERAL MOVEMENT

9-7. Lateral movement is moving at an angle to the left or right. During lateral movement—
   1. Roll the foot heel to toe to best provide a stable firing platform.
   2. Shoot while maintaining your natural walking gait.
   3. Keep the weapon ready.
   4. Maintain situational awareness.
   5. When moving, bring the weapon back slightly toward your body to aid in recoil and control.
   6. Maintain an aggressive posture.
   7. Walk in a straight line, keeping your center of gravity low (drop your hips).

Chapter 9

8. Keep the muzzle of the weapon downrange toward the target.
9. Use the upper body as a turret.
10. Twist at the waist.
11. Maintain proper posture.
12. Do not overstep or cross your feet, because this can decrease your balance and center of gravity.

TURNING MOVEMENT

9-8. Turning movement is used to engage widely dispersed targets in the oblique and on the flanks. Turning skills are just as valuable in a rapidly changing combat environment as firing on the move (such as lateral movement) skills are and should only be used with the alert carry.

Note. It does not matter which direction you are turning or which side is your strong side. You must maintain the weapon at an exaggerated low-alert carry for the duration of the turn.

9-9. When executing a turn to either side, the Soldier will—
1. Look first. Turn head into the direction of the turn first.
2. Orient the hips to the target as soon as possible.
3. Once you have a clear path to the target, draw and present your weapon smoothly.
4. Follow with your body.
5. Maintain situational awareness.
6. Maintain proper posture.

Note. Pivoting may lead to over or under rotation toward the target, lengthening the process.

Appendix A
Ammunition

This appendix discusses the characteristics and capabilities of pistol ammunition.

SMALL ARMS AMMUNITION CARTRIDGE

A-1. Ammunition for use in the pistol is described as a cartridge. A small arms cartridge is an assembly consisting of a cartridge case, a primer, a quantity of propellant, and a bullet. The following terminology describes the general components of the pistol cartridge:

- Cartridge case. The cartridge case is a brass, rimless, center-fire case that provides a means to hold the other components of the cartridge.
- Propellant. The propellant (or powder) provides the energy to propel the projectile through the barrel and downrange towards a target through combustion.
- Primer. The primer is a small explosive charge that provides an ignition source for the propellant.
- Bullet. The bullet or projectile is the only component that travels to the target.

Note. Dummy cartridges are composed of a cartridge case and bullet, with no primer or propellant.

CARTRIDGE CASE

A-2. Service pistols use rimless cartridge cases, which have extraction grooves. These cases are designed to support center-fire operation, where the primer is located in a small well in the center of the cartridge case head.

A-3. Center-fire cases have a centrally located primer pocket in the base of the case, which separates the primer from the propellant in the cartridge case. These cases are designed to withstand pressures generated during firing and are used for all small arms except caliber .22.

A-4. All 9-mm ammunition uses the rimless cartridge case. The rim diameter is the same as the case body and has an extractor groove. This design allows for the stacking of multiple cartridges in a magazine.

A-5. When the round is fired, the cartridge case assists in containing the burning propellant. The gases created expand the cartridge case tightly to the chamber walls to provide rear obturation.

PROPELLANT

A-6. Cartridges are loaded with various propellant weights that impart sufficient velocity, within safe pressure, to obtain the required ballistic projectile performance. The propellants are either a single-base (nitrocellulose) or double-base (nitrocellulose and nitroglycerine) composition.

A-7. The propellant may be a single-cylindrical or multiple-perforation, a ball, or a flake design to facilitate rapid burning. Most propellants are coated to assist the control of the combustion rate. A final graphite coating facilitates propellant flow and eliminates static electricity in loading the cartridge.

A-8. There are four different types of solid-propellant composition; single-based, double-based, triple-based, and composite. Small arms ammunition uses predominantly double-based solid propellants in military grade cartridges.

A-9. Double-based propellants consist of nitrocellulose with nitroglycerin or other liquid organic nitrate explosives added. Stabilizers and other additives are used to control the chemical stability and enhance the propellant's properties. Nitroglycerin reduces smoke and increases the chemical energy output of the propellant composition.

PRIMER

A-10. Center-fire small arms cartridges contain a percussion primer assembly. The assembly consists of a brass or gilding metal cup. The cup contains a pellet of sensitive explosive material secured by a paper disk and a brass anvil.

A-11. The weapon firing pin striking the center of the primer cup base compresses the primer composition between the cup and the anvil, which causes the composition to explode. Holes or vents located in the anvil or closure cup allow the flame to pass through the primer vent, igniting the propellant.

BULLET

A-12. The bullet is a cylindrically shaped lead or alloy projectile that engages the rifling of the barrel. The bullets used today are either lead (lead alloy), or assemblies of a jacket and lead. Some projectiles may be manufactured from plastic, wax, or plastic binder, and metal powder, two or more metal powders, or various combinations based on the cartridge's use.

SPECIFICATIONS

A-13. Figure A-1 shows the traits of the M882 ball round. For more information on data and characteristics for small-caliber ammunition, see TM 43-0001-27, *Army Ammunition Data Sheets for Small Caliber Ammunition (Federal Supply Class 1305) (Reprinted W/Basic INCL C1-13)*.

Ammunition

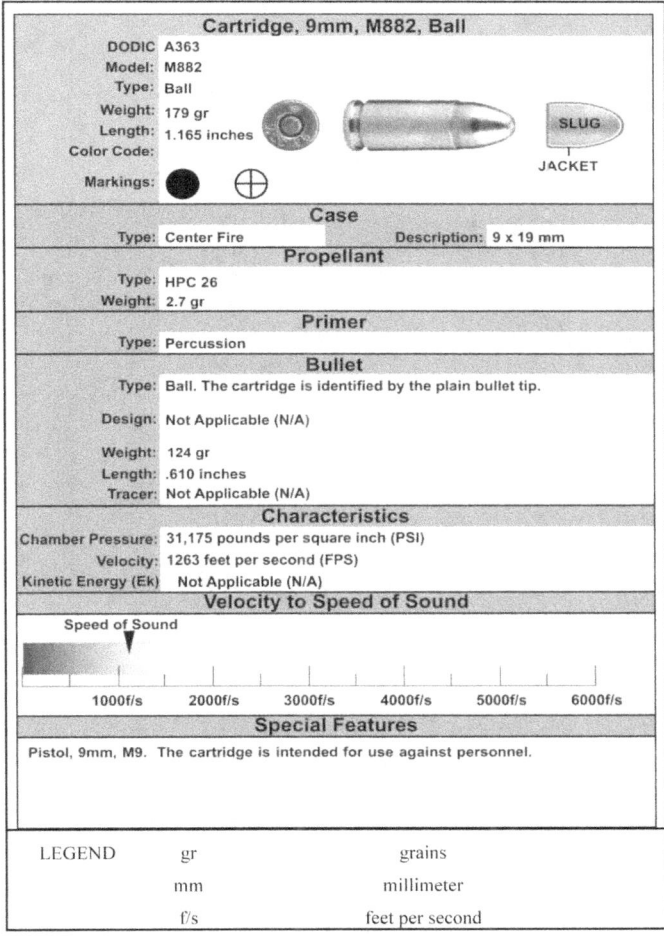

Figure A-1. M882 ball round

This page intentionally left blank.

Appendix B
Ballistics

Ballistics is the science of the processes that occur from the time a firearm is fired to the time when the bullet impacts its target. Soldiers must be familiar with the principles of ballistics as they are critical in understanding how the projectiles function, perform during flight, and the actions of the bullet when it strikes the intended target. The profession of arms requires Soldiers to understand their weapons, how they operate, their functioning, and their employment.

BALLISTICS CATEGORIES

B-1. The flight path of a bullet includes three stages: the travel down the barrel, the path through the air to the target, and the actions the bullet takes upon impact with the target. These stages are defined in separate categories of ballistics: internal, external, and terminal ballistics.

INTERNAL BALLISTICS

B-2. Internal ballistics is the study of the propulsion of a projectile. Internal ballistics begin from the time the firing pin strikes the primer to the time the bullet leaves the muzzle. Once the primer is struck the priming charge ignites the propellant. The expanding gases caused by the burning propellant create pressures which push the bullet down the barrel. The bullet engages the lands and grooves (rifling) imparting a spin on the bullet that facilitates stabilization of the projectile during flight. Internal ballistics ends at shot exit, where the bullet leaves the muzzle.

B-3. Several key terms are used when discussing the physical actions of internal ballistics —
- Bore. The interior portion of the barrel forward of the chamber.
- Chamber. The part of the barrel that accepts the ammunition for firing.
- Grain (gr). A unit of measurement of either a bullet or a projectile. There are 7000 grains in a pound, or 437.5 grains per ounce.
- Pressure. The force developed by the expanding gasses generated by the combustion (burning) of the propellant. Pressure is measure in pounds per square inch (psi).
- Shoulder. The area of the chamber that contains the shoulder, forcing the cartridge and projectile into the entrance of the bore at the throat of the barrel.

- Muzzle. The end of the barrel.
- Throat. The entrance to the barrel from the chamber. Where the projectile is introduced to the lands and grooves within the barrel.

EXTERNAL BALLISTICS

B-4. External ballistics is the study of the physical actions and effects of gravity, drag, and wind along the projectile's flight to the target. Exterior ballistics includes only those general physical actions that cause the greatest change to the flight of a projectile. External ballistics begins at shot exit and continues through the moment the projectile strikes the target.

B-5. Soldiers use the following terms and definitions to describe the actions or reactions of the projectile during flight. This terminology below is standard when dealing with any weapon or weapon system, regardless of caliber.

- Axis of the bore (line of bore). The line passing through the center of the bore or barrel.
- Line of sight (LOS) or gun target line. A straight line between the sights or optics and the target. The LOS is never the same as the axis of the bore. The LOS is what the Soldier sees through the sights and can be illustrated by drawing an imaginary line from the firer's eye through the rear and front sights out to infinity. The LOS is synonymous with the gun target line when viewing the relationship of the sights to a target.
- Line of elevation (LE). The angle from the ground to the axis of the bore.
- Ballistic trajectory. The path of a projectile when influenced only by external forces, such as gravity and atmospheric friction.
- Maximum ordinate. The maximum height the projectile travels above the line of sight on its path to the point of impact.
- Time of flight. The time taken for a specific projectile to reach a given distance after firing.
- Jump. Vertical jump in an upward and rearward direction caused by recoil.
- Line of departure (LD). The line the projectile is on at shot exit.
- Muzzle. The end of the barrel.
- Muzzle velocity or velocity. The velocity of the projectile measured at shot exit. Muzzle velocity decreases over time due to air resistance. For small arms ammunition, velocity (V) is represented in feet per second.
- Twist rate. The rotation of the projectile within the barrel of a rifled weapon based on the distance to complete one revolution. The twist rate relates to the ability to gyroscopically spin-stabilize a projectile on rifled barrels, improving its aerodynamic stability and accuracy. The twist rate of the M9 pistol is a right-hand, six groove (approximately one turn in ten inches, or R 1:10 inches).
- Shot exit. The moment the projectile clears the muzzle of the barrel, when the barrel no longer supports the bullet.

Ballistics

- Oscillation. The movement of the projectile in a circular pattern around its axis during flight.
- Drift. The lateral movement of a projectile during its flight caused by its rotation or spin.
- Yaw. A deviation from stable flight by oscillation. Cross wind or destabilization when the projectile enters or exits a transonic stage can cause yaw.
- Grain (gr). A unit of measurement of either a bullet or a propellant charge. There are 7000 grains in a pound, or 437.5 grains per ounce.
- Pressure. The force the expanding gases generate from the combustion (burning) of the propellant. For small arms, pressure is measured in pounds per square inch (psi).
- Gravity. The constant pressure of the earth on a projectile at a rate of about 9.8 meters per second squared, regardless of the projectile's weight, shape or velocity. Commonly referred to as bullet drop, gravity causes the projectile to drop from the line of departure.
- Drag (air resistance). The friction that slows the projectile down while moving through the air. Drag begins immediately upon the projectile exiting the barrel (shot exit). Drag slows the projectile's velocity over time, and is most pronounced at extended ranges. Each round has a ballistic coefficient that is a measurement of the projectile's ability to minimize the effects of air resistance (drag) during flight.
- Trajectory. The path of flight that the projectile takes upon shot exit over time. For the purposes of this manual, the trajectory ends at the point of impact.
- Wind. Wind has the greatest variable effect on ballistic trajectories. The effects of wind on a projectile are most noticeable in three key areas between half and two-thirds the distance to the target:
  - Time (T). The amount of time the projectile is exposed to the wind along the trajectory. The greater the range to target, the greater time the projectile is exposed to the wind's effects.
  - Direction. The direction of the wind in relation to the axis of the bore. The relation compensates for direction of the projectile's drift.
  - Velocity (V). The speed of the wind during the projectile's trajectory to the target. Variables in the overall wind velocity affecting a change to the ballistic trajectory include sustained rate of the wind and gust spikes in velocity.

Note. Wind isn't applicable to the pistol due to the relative short range engagements with a pistol.

TERMINAL BALLISTICS

B-6. Terminal ballistics is the science of the actions of a projectile from the time it strikes an object until it comes to rest (called terminal rest). This includes the terminal effects that take place against the target.

- Kinetic energy (KE). A unit of measurement of the delivered force of a projectile. Kinetic energy is the delivered energy that a projectile possesses due to its mass and velocity at the time of impact. Kinetic energy is directly related to the penetration capability of a projectile against the target.
- Penetration. The ability or act of a projectile to enter a target's mass based on its delivered kinetic energy. When a projectile strikes a target, the level of penetration into the target is termed the impact depth. The impact depth is the distance from the point of impact to the moment the projectile stops at its terminal resting place. Ultimately, the projectile stops when it has transferred its momentum to an equal mass of the medium (or arresting medium).

B-7. Against any target, penetration is the most important terminal ballistic consideration. Soldiers must be aware of the penetration capabilities of their ammunition against their target, and the most probable results of the terminal ballistics.

ACTIONS AFTER TRIGGER SQUEEZE

B-8. Once the trigger is squeezed, the ballistic action begins. Although not all ammunition and weapons operate in the same manner, the following list describes the general events that occur on the M9 pistol when the trigger is squeezed.

- The hammer strikes the rear of the firing pin.
- The firing pin is pushed forward, striking the cartridge percussion primer assembly.
- The primer is crushed, pushing the primer composition through the paper disk, and on to the anvil, detonating the primer composition.
- The burning primer composition is focused evenly through the primer cup vent hole, igniting the propellant.
- The propellant burns evenly within the cartridge case.
- The cartridge case wall expand from the pressure of the burning propellant, firmly locking the case to the chamber walls.
- The expanded cartridge case, held firmly in place by the chamber walls and the face of the bolt provide rear obturation, keeping the burning propellant and created expanding gasses in front of the cartridge case.
- The projectile is forced by the expanding gasses firmly into the lands and grooves at the throat of the bore, causing engraving.
- Engraving causes the scoring of the softer outer jacket of the projectile with the lands and grooves of the bore. This allows the projectile to spin at the twist rate of the lands and grooves, and provides a forward obturation seal. The forward obturation keeps the expanding gasses behind the projectile in order to push it down the length of the barrel.

Ballistics

- As the propellant continues to burn, the gasses created continue to seek the path of least resistance. As the cartridge case is firmly seated and the projectile is moveable, the gas continues to exert its force on the projectile.
- As the end of the projectile leaves the muzzle, it is no longer supported by the barrel itself. Shot exit occurs.
- Upon shot exit, most of the expanding and burning gasses move outward and around the projectile, causing the muzzle flash.
- At shot exit, the projectile achieves its maximum muzzle velocity. From shot exit until the projectile impacts an object, the projectile loses velocity at a steady rate due to air resistance.
- As the round travels along its trajectory, the bullet drops consistently by the effects of gravity.
- As the actual line of departure is an elevated angle from the line of sight, the projectile appears to rise and then descend. This rise and fall of the projectile is the trajectory.
- The round achieves the highest point of its trajectory typically over half way to the target, depending on the range to target. The high point is called the round's maximum ordinate (max ord).
- From the maximum ordinate, the projectile descends into the target.
- The round strikes the target at the point of impact, which, depending on the firing event, may or may not be the desired point of impact, and is seldom the point of aim.
- Once the projectile strikes a target or object, it delivers its kinetic energy (force) at the point of impact.

TERMINAL BALLISTICS BEGIN

B-9. Once terminal ballistics begin, no bullets follow the same path or function. Generally speaking, each projectile penetrates objects where the delivered energy (mass times velocity squared, divided by 2) is greater than the mass, density, and area of the target at the point of the delivered force. Other factors, such as the angle of attack, yaw, and oscillation, and other physical considerations are not included in this ballistic discussion.

SOFT TISSUE PENETRATION

B-10. A gunshot wound, or ballistic trauma, is a form of physical damage sustained from the entry of a projectile. The degree of tissue disruption caused by a projectile is related to the size of the cavities the projectile creates as it passes through the target's tissue. When striking a personnel target, there are two types of cavities created by the projectile: permanent and temporary wound cavities.

Permanent Wound Cavity

B-11. The permanent cavity refers specifically to the physical hole left in the tissues of soft targets by the pass-through of a projectile. It is the total volume of tissue crushed or destroyed along the path of the projectile within the soft target.

B-12. Depending on the soft tissue composition and density, the tissues are either elastic or rigid. Elastic organs stretch when penetrated, leaving a smaller wound cavity. Organs that contain dense tissue, water, or blood are rigid, and can shatter from the force of the projectile. When a rigid organ shatters from a penetrating bullet, it causes massive blood loss within a larger permanent wound cavity. Although typically fatal, striking these organs may not immediately incapacitate the target.

Temporary Wound Cavity

B-13. The temporary wound cavity is an area that surrounds the permanent wound cavity. It is created by soft, elastic tissues as the projectile passes through the tissue at greater than 2000 feet per second. The tissue around the permanent cavity is propelled outward (stretched) in an almost explosive manner from the path of the bullet. This forms a temporary recess or cavity 10 to 12 times the bullet's diameter.

B-14. Tissue such as muscle, some organs, and blood vessels are elastic and can be stretched by the temporary cavity with little or no damage and have a tendency to absorb the projectile's energy. The temporary cavity created slowly reduces in size over time, although typically not returning completely to the original position or location.

Note. Projectiles that do not exceed 2000 feet per second velocity on impact do not provide sufficient force to cause a temporary cavity capable of incapacitating a threat.

B-15. The extent of the cavitation (the bullet's creation of the permanent and temporary cavities) is related to the characteristics of the projectile:
- Kinetic energy (KE). The delivered mass at a given velocity. Higher delivered kinetic energy produces greater penetration and tissue damage.
- Yaw. Any yaw at the point of impact increases the projectiles surface area that strikes the target, decreasing kinetic energy, but increasing the penetration and cavity size.
- Deformation. The physical changes of the projectile's original shape and design due to the impact of the target. Deformation increases the projectile's surface area and the size of the cavity created after penetration.
- Fragmentation. The fracturing of a projectile into multiple pieces or subprojectiles. The multiple paths of the fragmented subprojectiles are unpredictable in size, velocity, and direction. The bullet jacket, and for some types of projectiles, the lead core, fracture creating small, jagged, sharp edged pieces that are propelled outward with the temporary cavity. Fragments can sever tissue and causing large explosive-type damage to the body. Bone fragments caused by the bullet's strike can have the same effect.

Ballistics

- Tumbling. Tumbling is the inadvertent end-over-end rotation of the projectile. As a projectile tumbles as it strikes the target, the bullet travels through the tissues with a larger diameter. Tumbling causes a more severe permanent cavity as it passes through the soft tissue. A tumbling projectile can change direction erratically within the body due to its velocity and tendency to strike dense material with a larger surface area.

B-16. Once inside the target, the projectile's purpose is to destroy soft tissues with fragmentation. The ball ammunition is designed to not flatten or expand on impact, which would decrease velocity and delivered energy.

Incapacitation

B-17. Incapacitation with direct fire is the act of ballistically depriving a target of the ability, strength, or capability to continue its tactical mission. To assist in achieving the highest probability of incapacitation with a single shot, the projectile is designed with the ability to tumble, ricochet, or fragment after impact.

B-18. The projectile or its fragments then must hit a vital, blood-bearing organ or the central nervous system to effectively incapacitate the threat. The projectile's limited fragmentation potential after entry maximizes the soft tissue damage and increases the potential for rapid incapacitation.

Lethal Zones

B-19. The Soldier's primary point of aim at any target by defau t is center of visible mass, which allows for a tolerance that includes the greatest margin of error with the highest probability of a first-round hit. The combat conditions may require more precise fires at partially exposed targets or targets that require immediate incapacitation.

B-20. Ideally, the point of aim is anywhere within a primary switch area. This point maximizes the possibility of striking major organs and vessels, rendering a clean, one-shot kill (see figure B-1 on page B-8).

Appendix B

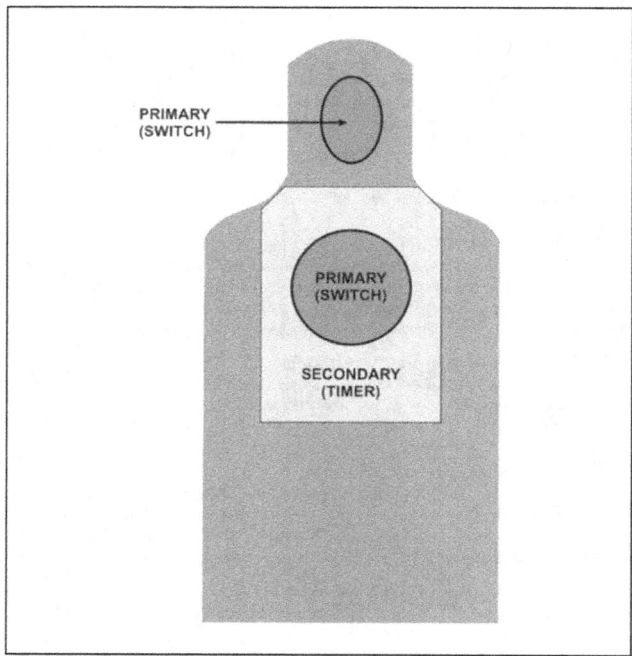

Figure B-1. Lethal zone example

B-21. Shots to the head should be weighed with caution. The head is the most frequently moved body part and is the most difficult to hit with precision. Soldiers should consider shots to other exposed body parts, such as the pelvic area.

B-22. Shots to the pelvic area are used when the target is not completely visible or when the target is wearing body armor that prevents the Soldier from engaging the primary zone. This area is rich in large blood vessels and a shot here has a good possibility of impeding enemy movement by destroying the pelvis or hitting the lower spine.
- Circuitry shots (switches).
- Hydraulic shots (timers).

Ballistics

Circuitry Shots (Switches)

B-23. Circuitry shots, or switches, are strikes to a target that deliver its immediate incapacitation. Immediate incapacitation is the sudden physical or mental inability to initiate or complete any physical task. To accomplish this, the central nervous system must be destroyed by hitting the brain or spinal column. All bodily functions and voluntary actions cease when the brain is destroyed and if the spinal column is broken, all functions cease below the break.

Hydraulic Shots (Timer)

B-24. Hydraulic shots, or timers, are impacts on a target where immediate incapacitation is not guaranteed. These types of ballistic trauma are termed timers because after the bullet strikes, the damage caused requires time for the threat to have sufficient blood loss to render them incapacitated. Hydraulic shots, although ultimately lethal, allow for the threat to function in a reduced capacity for a period of time.

Note. For hydraulic shots to eliminate the threat, they must cause a 40 percent loss of blood within the circulatory system. If the shots do not disrupt that flow at a rapid pace, the target can continue their mission. Once two liters of blood are lost, the target transitions into hypovolemic shock and becomes incapacitated.

This page intentionally left blank.

Appendix C
Complex Engagements

This appendix provides detailed information on complex engagements with the service pistol to include limited visibility engagements, rifle to pistol transitions, engaging a moving target, and chemical, biological, radiological, and nuclear (CBRN) operations.

LIMITED VISIBILITY

C-1. Image intensifiers, IR pointers, and thermal optics can be used in conjunction with a pistol. The Soldier can be wearing the AN/PSQ-20 enhanced night vision goggle and using the AN/PEQ-14 at the same time.

SECONDARY WEAPON LIGHT

C-2. A weapon-mounted light is required and can enhance a Soldier's lethality in conducting low-light operations with a pistol. (See figure C-1.) Not only is this beneficial in an urban environment, but also in subterranean environments (tunnels). Lights can disorient or blind a target temporarily, or it can be used as a control technique. To search and scan a room or tunnel, Soldiers should use short, sweeping motions in tandem with other members of the team.

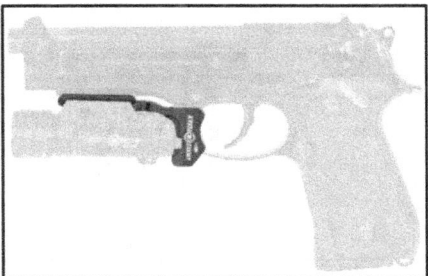

Figure C-1. Weapon mount and flashlight

- Ensure the holster has a secure cap on the bottom so if the light goes off accidentally you are not compromised.
- Soldiers should focus their beam prior to mission, too wide of a beam can prevent Soldiers from seeing details as clearly as possible.

Appendix C

- Protect the flashlight lens from carbon buildup by smearing it with a waxy substance such as the lip balm that is readily available through supply channels.

Note. Install the lights on both your primary and secondary weapons so you can operate either light with either hand.

PRIMARY WEAPON LIGHT

C-3. Even if the primary weapon malfunctions, the Soldier can still use the light source on that weapon. The Soldier would support the primary weapon with the nonfiring arm. The Soldier would aim the light source from the primary weapon at the target, and then draw, present, and fire the secondary weapon using the single-hand grip.

PRIMARY TO SECONDARY TRANSITION

C-4. In close quarters, if the Soldier experiences any of the following malfunctions:
- Light recoil.
- Dead trigger.
- Click, no bang.

C-5. Then the Soldier--
1. Tries to put the rifle on safe.
2. Moves firing hand to the pistol.
3. Indexes pistol.
4. Using supporting hand, moves primary weapon down and out of the way.

Note. Ensure location of primary weapon does not block access to secondary weapon.

5. With the firing hand, draws and presents the pistol towards target.
6. With supporting hand, meets and greets the pistol.
7. Finishes presentation.

C-6. After the engagement, because the primary weapon experienced a malfunction, the Soldier may use the pistol light to observe the chamber of the primary weapon. If clear, the Soldier places the primary weapon back into operation, then reloads, makes safe, and holsters the secondary weapon.

Note. Within 25 meters of the target, if you have already experienced a malfunction with the primary weapon, transition to the secondary weapon is recommended.

Complex Engagements

MOVING TARGETS

C-7. Moving targets are those threats that appear to have a consistent pace and direction. Targets on any battlefield do not remain stationary for long periods, particularly once a firefight begins. Soldiers must have the ability to deliver lethal fires at a variety of moving target types and be comfortable and confident in the engagement techniques. There are two methods for defeating moving targets; tracking and trapping.

TRACKING METHOD

C-8. The tracking method is used for a moving target that is progressing at a steady pace over a well-determined route. If a Soldier uses the tracking method, he or she tracks the target with the pistol's sights while maintaining a point of aim on or ahead of (leading) the target until the shot is fired. (See figure C-2.)

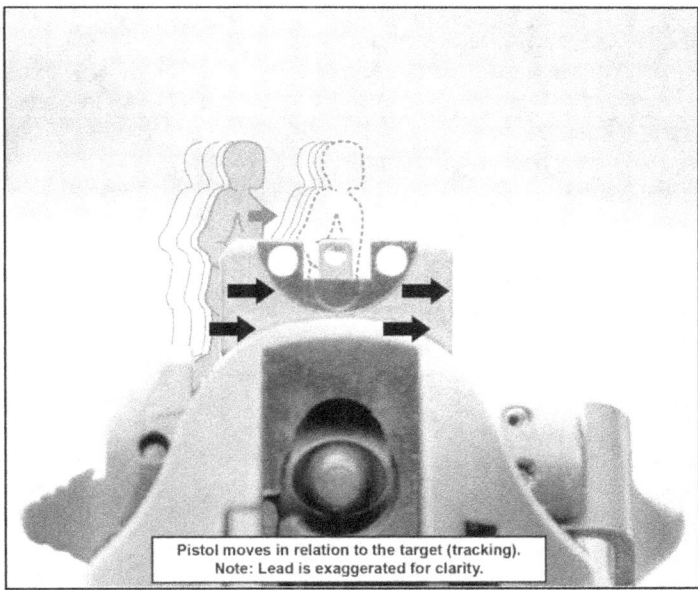

Figure C-2. Tracking method

Note. When establishing a lead on a moving target, the pistol sights are not centered on the target, but held on a lead in front of the target instead.

Appendix C

TRAPPING METHOD

C-9. The trapping method (see figure C-3) is used when it is difficult to track the target with the aiming device, as in the prone or sitting position. The lead required to effectively engage the target determines the engagement point and the appropriate hold off.

C-10. With the sights settled, the target moves into the predetermined engagement point and creates the desired sight picture. The trigger is pulled simultaneously with the establishment of sight picture. To execute the trapping method, a Soldier performs the following steps:

- Select an aiming point ahead of the target, which is where to set the trap.
- Obtain sight alignment on the aiming point.
- Hold sight alignment until the target moves into vision and the desired sight picture is established.
- Engage the target once sight picture is acquired.
- Follow-through so the pistol sights are not disturbed as the bullet exits the muzzle.

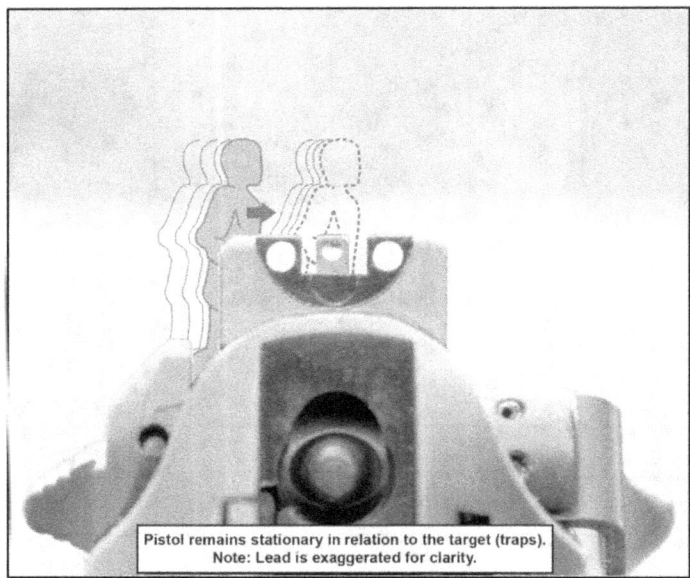

Figure C-3. Trapping method

OBLIQUE TARGETS

C-11. Threats that are moving diagonally toward or away from the Soldier are oblique targets. They offer a unique problem set to Soldiers where the target may be moving at a steady pace and direction; however, their oblique posture makes them appear to move slower.

EVASIVE TARGETS

C-12. Moving targets that change speed and direction, such as a tactically moving threat, are classified as evasive targets. They provide limited opportunity for the Soldier to apply the correct hold off to achieve a target hit.

LIMITED EXPOSURE TARGETS

C-13. Threats moving tactically present a limited exposure condition that may reduce the viewable size of a threat or cause targets to appear in and out of the sight picture after they are initially acquired. In these situations, Soldiers may choose to apply a hold off for where a target is estimated to present itself next, rather than wait for the target to present itself for a more refined point of aim.

CHEMICAL OPERATIONS

C-14. When firing a pistol in CBRN conditions, the Soldier should use optical inserts, if applicable. Firing in mission-oriented protective posture (MOPP) levels 1 and 2 is seldom a problem for a Soldier. Unlike when wearing a protective mask while firing a pistol, a Soldier can acquire a sight picture the same as he or she would while not wearing a protective mask.

C-15. However, firing in mission-oriented protective posture levels 3 and 4 might be complicated by inclusion of the mask and gloves that are worn during this level. The use of gloves may require the Soldier to adjust for proper grip and trigger control. The use of the protective mask may require the Soldier to make minor adjustment to his or her position to acquire the correct sight alignment and sight picture.

Note. Soldiers should practice firing in all mission-oriented protective posture levels to become proficient in pistol CBRN firing.

This page intentionally left blank.

Appendix D
Drills

This appendix describes drills for the pistol that enforce and enhance gun handling skills needed to succeed in Tables IV through VI of the Integrated Weapon Training Strategy.

The drill structure is standardized for all individual weapons to reinforce the most common actions all Soldiers routinely execute with their assigned equipment.

These drills are used during Table III of the integrated weapons training strategy, as well as during routine maintenance, concurrent training, and during deployments. The drills found within this appendix are used to build and maintain skills needed to achieve proficiency and mastery of the weapon, and are to be ingrained into daily use with the weapon.

Note. The drills listed in this appendix are for DRY-FIRE purposes only. Units may add to the drills that are listed here and are encouraged to develop additional drills to include pistol LIVE-FIRE drills that they can use to augment with the Army service pistol training strategy.

BUILDING CONFIDENCE

D-1. Each drill is designed to develop confidence in the equipment and Soldier actions during training and combat operations. As they are reinforced through repetition, they become second nature to the Soldier, providing smooth, consistent employment during normal and unusual conditions. The drills provided are designed to build the Soldier's proficiency with the following principles:

- Mindset. Perform tasks quickly and effectively under stress.
- Efficiency. Ensure the drills require the least amount of movement or steps to complete correctly. Make every step count.
- Individual tactics. Ensure the drills are directly linked to employment in combat.
- Flexibility. Provide drills that are not rigid in execution.

Mindset

D-2. Continuous combat is inherently stressful. It exhausts you and causes physiological changes that reduce your ability to perform tasks as quickly or effectively as necessary. Your ability to function under stress is the key to winning battles, since, without you, weapons and tactics are useless. Individual and unit military effectiveness depend on your ability to think clearly, accurately, quickly, all with initiative, motivation, physical strength, and endurance.

D-3. The impact of physiological changes caused by the stress of combat escalates or deescalates based on the degree of stimulation, causing you to attain different levels of awareness as events occur in the continually transitioning operational area around you. Maintaining a tactical mindset involves understanding and transitioning between levels of awareness as the situation requires escalation or de-escalation.

Note. Stress can be countered using the principles associated with Soldier resilience and performance enhancement. The Comprehensive Soldier and Family Fitness (CSF2) program is designed to increase a Soldier's ability and willingness to perform an assigned task or mission and enhance performance by assessing and training mental resilience, physical resilience, and performance enhancement techniques and skills. This initiative introduces many resources used to train Soldiers on skills to counter stress.

Efficiency

D-4. Efficiency is defined as using the least resources to achieve an end. Efficient movements are fast because they include only necessary actions. This reduces mental and physical effort, allowing you to repeat the movement until you can predict the effect. This lets you focus on tactics while still producing accurate and precise fires.

Individual Tactics

D-5. Individual tactics are actions independent of unit standard operating procedures or situations that maximize your chance of survival and victory in a small arms direct fire battle. Examples of individual tactics include cover and standoff, or manipulation of time and space between your and his or her enemy.

Flexibility

D-6. These techniques are not prescriptive; multiple techniques can be used to achieve the same goal. In fact, there is no single way to use a pistol; different types of enemies and scenarios require different techniques. However, the techniques presented are efficient and proven techniques for conducting various pistol-related tasks. Should other techniques be selected, they should meet the following criteria:

Reliable Under Conditions of Stress

D-7. Techniques should be designed for reliability when it counts; during combat. The technique should produce the intended results without fail, under any conditions and while wearing mission-essential equipment. It should also be tested under as high stress conditions as allowed in training.

Repeatable Under Conditions of Stress

D-8. As combat is a stressor, a Soldier's body responds much as it does to any other stressful stimulus; physiological changes begin to occur, igniting a variable scale of controllable and uncontrollable responses based on the degree of stimulation. The technique should support or exploit the body's natural reaction to life-threatening stress.

Efficient in Motion

D-9. The technique should create efficiency. It should contain only necessary movement. Extra movement wastes time. Time can make the difference between life and death. Consider how fast violent encounters occur; an unarmed person can cover 20 feet in about a second.

Develop Natural Responses Through Repetition

D-10. When practiced correctly and in sufficient volume, the technique should build reflexive reactions that a Soldier applies in response to a set of conditions. Only with correct practice will a Soldier create the muscle memory necessary to serve them under conditions of dire stress. The goal is to create automaticity, the ability to perform an action without thinking through the steps associated with the action.

Leverage Overmatch Capabilities

D-11. Fast and efficient presentation of the pistol leaves more time to refine aim. You will develop speed throughout the training cycle and maintain it during operations.

D-12. The sooner you can get your weapon pointed at the target, the more time you have to refine your aim. The closer the target gets, the less time you have to refine your aim. You have to make every second count.

CONDUCT DRILLS

D-13. To build the skills necessary to master the functional elements of the shot process, certain tasks are integrated into drills. These drills are designed specifically to capture the routine, critical tasks or actions Soldiers must perform fluently and as a second nature to achieve a high level of proficiency.

D-14. Drills focus on the Soldier's ability to apply specific weapons manipulation techniques to engage a threat correctly, overcome malfunctions of the weapon or system, and execute common tasks smoothly and confidently.

Drill A: Weapon Check

D-15. Visually inspect the weapon on first receiving it from the arms room or storage facility. Check the weapon to ensure, at least, that—
- Weapon is clear.
- Pistol functions correctly.
- Magazines are serviceable.

Drill B: Equipment Check

D-16. Before combat, check to be sure your equipment and accessories are ready for operation. That is, make sure—
- Batteries are charged.
- Holster is secured and placed correctly.
- Holster does not interfere with tactical movement.
- Basic load of magazines are stowed properly.

Drill C: Load and Conduct Status Check

D-17. Remember to load and conduct status check on your secondary weapon first—then load and conduct status check on primary weapon.

Command: LOAD AND CONDUCT STATUS CHECK.

1. **Load.**
    a. Draw the pistol with your trigger finger pressed straight along the frame of the pistol.
    b. Bring the pistol to your workspace.
    c. Ensure the decocker is up (M9 only).
    d. Rotate the pistol to the target.
    e. Pull back on the rear serrations while placing upward pressure on the slide stop/slide release.
    f. Inspect the chamber, magazine well and bolt face for debris and remaining rounds.
    g. Withdraw magazine from a pouch with finger toward bullets.
    h. Place the pistol and the magazine on the same angle and index magazine into magazine well with force.
    i. Rotate the pistol to target and release the slide with the support thumb.

2. **Conduct Status Check.**
    j. Return the pistol to your workspace.
    k. Grasp the rear sight with the index finger of your supporting hand and grasp the back strap with the thumb of your supporting hand.
    l. Pull the slide back slightly to reveal the round casing.
    m. Let go of the slide and return to locking. If needed, tap back of slide to ensure slide is fully forward.

Appendix D

DRILL D: DRAW/HOLSTER FROM DIFFERENT CARRY POSITIONS

D-18. Draw and present the pistol.

    Command: DRAW AND PRESENT.

1. **Draw.**
    a. Assume a correct stance with feet with your feet about shoulder width apart, knees slightly bent, and your weight on the balls of your feet.
    b. Start with your hands relaxed, moving both hands simultaneously.
    c. Move supporting hand to indexed position (center of chest, bottom of pectoral).
    d. With your firing hand, index the pistol.
    e. Seat your firing hand firmly and high on the pistol grip.
    f. Defeat retention of the holster.
    g. Draw the pistol straight up and rotate.
    h. Keep your trigger finger straight and pressed along the frame.
    i. Move your firing hand forward of the support hand, and allow your supporting hand to index the pistol.
    j. With your supporting hand, firmly grasp the pistol grip.

2. **Present.**
    k. Finish the grip as you present the pistol to the target.
    l. Ensure the thumb of your firing hand is on top of the thumb of your supporting hand. Lock out the wrist of the supporting hand.
    m. Prepare the trigger as you push out. Let your trigger finger contact the trigger when you see the sights and target in the same field of view.
    n. At full presentation, maintain a high, firm grip, correct sight alignment and sight picture, and proper stance. Keep elbows slightly bent to assist with recoil.

Note. The draw should lead to the sights being generally on target without any thought.

Drills

DRILL E: FIGHT DOWN

D-19. Draw and present the pistol in the standing position. On command, holster and move to the kneeling position. Assume the kneeling position, and then draw and present the pistol to the target. Holster the pistol before moving to the prone position. Repeat each position at least of three times.

Note. Leaders conduct this drill with Drill F, which is the same, but in reverse.

DRILL F: FIGHT UP

D-20. Draw and present the pistol in the prone position. On command, holster and move to the kneeling position. Assume the kneeling position, and then draw and present the pistol to the target. Holster the pistol before moving to the standing position. Repeat each position at least of three times.

Note. Leaders conduct this drill with Drill E, which is the same, but in reverse.

DRILL G: TRANSITION TO SECONDARY

D-21. Transitions to your secondary weapon when two conditions occur at the same time: you engage targets within 25 meters with your primary weapon, *and* your primary weapon malfunctions.

1. Try to put the rifle on safe.
2. Move firing hand to the pistol to defeat holster retention and establish a high, firm grip on the pistol.
3. Using supporting hand, move rifle out of the way.

Note. Ensure rifle is out of the way of pistol magazines and does not impede the draw.

4. With firing hand, draw pistol.
5. Move supporting hand to meet and greet position.
6. Present pistol.
7. Once you have presented the pistol and neutralized the target, holster the pistol and clear the malfunction of the rifle.
8. After the engagement, because the primary weapon experienced a malfunction, you may use the pistol light to observe the chamber of the primary weapon. If clear, place the weapon back into operation, then reload, make safe, and holster the secondary weapon.

Appendix D

DRILL H: RELOAD

D-22. The two types of reloads for pistol are reload and reload with retention.

Reload

D-23. You have used a whole magazine and the slide has locked to the rear. Seek cover and replace the magazine with a full one:
1. Keep eyes downrange and body oriented towards threat area.
2. Keep trigger finger straight and pressed along the frame.
3. Return the pistol to the workspace.
4. With supporting hand, begin magazine sweep, starting at the navel.
5. Regrip pistol to eject magazine.
6. Eject the magazine with the pistol vertical.

Note. In your peripheral vision, you should see the magazine fall.

7. Restore normal grip.
8. Shift focus to pistol.
9. With supporting hand, index magazine into the magazine well.
10. Shift focus back to target.
11. Depress slide release with supporting thumb (left handed shooters can come under frame and depress with index finger) with pistol pointed downrange.
12. Present pistol to target.

Note. If on a knee, look over shoulder (to the rear) before coming off knee.

Reload with Retention

D-24. You have used part of a magazine. Seek cover and replace with a full magazine and try to stow the partially full magazine:
1. Keep eyes downrange and body oriented towards threat area.
2. Keep trigger finger straight and pressed along the frame.
3. Return the pistol to the workspace.
4. With supporting hand, begin magazine sweep, starting at the navel.
5. Regrip pistol to eject magazine.
6. Eject the magazine with the pistol vertical.
7. Exchange fresh magazine with partial magazine using finger switch technique.

Drills

8. Restore normal grip.
9. Shift focus to pistol.
10. Make one attempt to stow partial magazine.
11. Shift focus back to target.
12. Present pistol to target.

Note. If on a knee, look over shoulder (to the rear) before coming off knee.

DRILL I: CLEAR MALFUNCTION

D-25. Rapidly clear the most common malfunctions on a pistol. At the same time, maintain muzzle and situational awareness. The best way to clear a malfunction of a pistol is immediate action.

Immediate Action

D-26. If your pistol experiences a failure to fire, conduct tap, rack, bang:
   1. Remove trigger finger from the trigger and place it straight alongside the frame.
   2. Bring pistol back into workspace.
   3. Rotate pistol.
   4. With the heel of the nonsupporting hand, forcefully tap up against the base of the magazine.
   5. Rotate the pistol to observe chamber and rack the slide.

Note. Allow the chambered round to extract and fall to the floor. (Observe chambered round).

   6. Rotate pistol back to target and reengage bang.

Remedial Action

   1. Start at tap, rack.
   2. Recognize the issue in the chamber.
   3. Lock slide to the rear.
   4. Drop magazine and allow gravity to work out the round
   5. Check chamber.
   6. Reinsert magazine or seat new magazine.

Appendix D

DRILL J: UNLOAD/SHOW CLEAR

D-27. When unloading, always unload and show clear with whatever weapon is in your hands. If your rifle is loaded and in your hands, unload and show the rifle clear first, then transition to pistol and unload and show it clear second.

> *Note.* This drill can be executed without ammunition in the weapon. Leaders may opt to use dummy ammunition.

Command: UNLOAD AND SHOW CLEAR.

1. Draw the pistol with your finger pressed straight along the frame of the pistol.
2. Bring the pistol to your workspace.
3. Ensure the decocker is on black (M9 only).
4. Place the pistol into the workspace and eject the magazine with the pistol vertical.
5. Stow the magazine.
6. Rotate the pistol to target.
7. Pull back on the slide while placing upward pressure on the slide stop/slide release.
8. Allow the chambered round to fall to the floor. (Muzzle control)
9. Inspect the chamber, magazine well, and bolt face for debris and remaining rounds.
10. Have teammate conduct a three-point safety check.
11. Release the slide and reholster (look the pistol back into the holster).

Appendix E
Qualification

This appendix explains the Combat Pistol Qualification Course (CPQC). If it is unavailable, the Alternate Pistol Qualification Course (APQC) may be used to sustain training and to qualify firers. DA Form 5704, *Alternate Pistol Qualification Course Scorecard* may be downloaded from the Army Publishing Directorate website.

The tower operator is completely responsible for and in charge of the range and the course. He or she controls absolutely all activities related to firing. Only the tower operator may issue the order to fire. Scorers and firers must await the tower operator's orders.

COURSE INFORMATION

E-1. The CPQC (shown in TC 25-8, *Training Ranges*) requires the Soldier to engage single and multiple targets at various ranges using the fundamentals of quick fire.

EXTRA ROUNDS

E-2. For each table of the CPQC, the firer is given extra rounds to reengage missed targets. Although only 30 targets will be exposed during the entire course, each firer will receive 40 rounds of ammunition. Hitting a target with an additional round during the exposure time is just as effective as hitting it with the first round. Consequently, the firer is not penalized for using or not using the extra ammunition. However, any unused ammunition must be turned in at the end of the table, and may not be used in any other table.

MAGAZINE CHANGES

E-3. Only three magazine changes are required during this course: one change in Firing Table II, and two changes in Firing Table V. For safety, each of these two tables begins with a magazine loaded only with one round. The first target appears, and the firer engages it with that round. By the time another target appears eight seconds later, the firer must have reloaded and prepared to engage. He or she will receive no commands to reload.

E-4. Failure to reload in time to engage the second target is scored as a miss. This teaches the Soldier to change magazines instinctively, quickly, and safely under pressure. In Table V, a second magazine change is commanded by the control tower.

Double-Action Mode

E-5. When firing the 9-mm pistol, the Soldier uses double action to fire the first round in every table.

Range to Target

E-6. The range to exposed targets must not exceed 31 meters from the firer. Table E-1 shows target exposure times for each firing table.

Table E-1. Target-exposure times

TARGET EXPOSURE TIMES			
SINGLE TARGETS	3 seconds	2 seconds	10 seconds
MULTIPLE TARGETS	5 seconds	4 seconds	20 seconds

STANDARDS BY FIRING TABLE

E-7. All qualification tables apply for day, night, and CBRN qualification. Each are described below. The standing firing position is used throughout the qualification.

Notes. 1. The range officer in charge determines a common target sequence for all lanes. This keeps a firer from getting ahead of adjacent firers.

2. Target sequences vary in distance from the firer, starting with no more than two targets at 10 meters and WITH the farthest targets at 31 meters.

3. The firer will remain in the same firing lane throughout the CPQC.

TABLE I--DAY STANDING

E-8. For this table, the firer receives one magazine with seven rounds in it. Five targets (single) are exposed. The firer assumes the standing firing position at the firing line and holds the weapon at the ready. The tower operator sets the target sequence.

TABLE II--DAY STANDING

E-9. For this table, the firer receives two magazines: one containing one round, and the other containing seven rounds. Six targets (four single and one set of two) are exposed:

First Magazine

E-10. The firer loads the first magazine (containing one round). One target is exposed.

Qualification

Second Magazine

E-11. After firing the round in the first magazine, the firer must change magazines at once. The firer has eight seconds to load the second magazine (containing seven rounds) and prepare to fire before the next target is exposed. Once it appears, the firer must engage in the three seconds before it is lowered. Failure to do so is scored as a miss.

TABLE III--DAY STANDING

E-12. For this table, the firer receives one magazine containing seven rounds. Five targets (three single and one set of two) are exposed.

TABLE IV--DAY STANDING

E-13. For this table, the firer receives one magazine containing five rounds. Four targets (two single and one set of two) are exposed.

TABLE V--DAY MOVING OUT

E-14. For this table, the firer receives three magazines: one each with one, seven, and five rounds. Ten targets are exposed. The firer begins 10 meter behind the firing line, in the middle of the trail. They take the following steps:

- The firer loads the first magazine (containing one round) and places the second magazine (containing seven rounds) in the magazine pouch closest to his or her firing hand. The firer places the third magazine (containing five rounds) in the magazine pouch farthest from his or her firing hand.
- (When the firer reaches the firing line, a single target is exposed. The firer has two seconds to hit it before it is lowered. Then has eight seconds to load the second magazine (containing seven rounds).
- At the end of eight seconds, another single target is exposed to the firer. If the firer has not loaded the second magazine in time to engage this target, this round is scored as a miss.
- When the tower operator is sure that the firing line has completed the magazine change, the tower operator commands MOVE OUT. Two multiple targets are the exposed, one after the other, at various ranges from the firer.
- After two sets of multiple targets are exposed, the Soldier is commanded to load the five-round magazine. After the command MOVE OUT is given, the remaining targets are presented to the firer in sequence. After the last targets are hit or lowered, the firer clears the weapon.
- The firer holds the weapon in the raised pistol position with the slide to the rear. The firer returns to the starting point and unloads and shows clear and reholsters the pistol. Any excess ammunition is turned in to the ammunition point. On hearing the order to do so, he or she moves to the firing line.

TABLE VI--DAY STANDING, CBRN

E-15. All firers will wear protective masks with hoods. For this table, the firer receives one magazine containing seven rounds. Five targets (three single and one set of two) are exposed.

TABLE VII--NIGHT STANDING

E-16. For this table, the firer receives one magazine containing five rounds. Four targets (two single and one set of two) are exposed.

Note. Commanders may use the Engagement Skills Trainer 2000 to conduct Firing Tables VI and VII (CBRN and night fire).

TOWER OPERATOR'S AUTHORITY

E-17. The tower operator is responsible for the range. For this reason, only the tower operator can give orders to scorers and firers on the range.

CONDUCT OF FIRE BY FIRING TABLE

E-18. For each table, the tower operator has scorers issue only the rounds required for that table. There are several fire commands to run range fire on the CPQC:

Table I--Day Standing

E-19. The tower operator orders firers to move to the firing line in preparation for firing. On command, the scorer issues to the firer one magazine containing 7 rounds. The tower operator commands—

TABLE ONE, STANDING POSITION, 7 ROUNDS.
LOAD AND CONDUCT STATUS CHECK (HOLSTER PISTOL).
READY ON THE RIGHT.
READY ON THE LEFT.
READY ON THE FIRING LINE.
MAKE READY (UNHOLSTER).
FIRERS, WATCH YOUR LANE.

E-20. The tower operator exposes the targets to the firers. When all targets have been exposed and engaged or lowered, the tower operator commands—

CEASE FIRE.
CLEAR ALL WEAPONS.
CLEAR ON THE RIGHT.
CLEAR ON THE LEFT.
THE FIRING LINE IS CLEAR.
UNLOAD AND SHOW CLEAR (HOLSTER PISTOL).
FIRERS AND SCORERS, MOVE DOWNRANGE AND CHECK YOUR TARGETS.
MARK AND COVER ALL HOLES.

Table II--Day Standing

E-21. The tower operator orders firers to secure their weapons. On command, the scorer issues to the firer one magazine containing a single round and another magazine containing seven rounds. The tower operator commands—

Qualification

TABLE TWO, STANDING POSITION, EIGHT ROUNDS.
LOAD AND CONDUCT STATUS CHECK ONE MAGAZINE WITH 1 ROUND (HOLSTER PISTOL).
LOAD YOUR 7-ROUND MAGAZINE WITHOUT COMMAND.
READY ON THE RIGHT.
READY ON THE LEFT.
READY ON THE FIRING LINE.
MAKE READY (UNHOLSTER PISTOL).
FIRERS, WATCH YOUR LANES.

E-22. The tower operator exposes the targets to the firers. When all targets have been exposed and engaged or lowered, the tower operator commands—

CEASE FIRE.
CLEAR ALL WEAPONS.
CLEAR ON THE RIGHT.
CLEAR ON THE LEFT.
THE FIRING LINE IS CLEAR.
UNLOAD AND SHOW CLEAR (HOLSTER PISTOL).
FIRERS AND SCORERS, MOVE DOWNRANGE AND CHECK YOUR TARGETS.
MARK AND COVER ALL HOLES.

Table III--Day Standing

E-23. The tower operator orders the firers to secure their weapons. On command, the scorer issues to the firer one magazine containing seven rounds. The tower operator commands—

TABLE THREE, STANDING POSITION, 7 ROUNDS.
LOAD AND CONDUCT STATUS CHECK (HOLSTER PISTOL).
READY ON THE RIGHT.
READY ON THE LEFT.
READY ON THE FIRING LINE.
MAKE READY (UNHOLSTER PISTOL).
FIRERS, WATCH YOUR LANES.

E-24. The tower operator exposes the targets to the firers. When all targets have been exposed and engaged or lowered, the tower operator commands—

CEASE FIRE.
CLEAR ALL WEAPONS.
CLEAR ON THE RIGHT.
CLEAR ON THE LEFT.
THE FIRING LINE IS CLEAR.
UNLOAD AND SHOW CLEAR (HOLSTER PISTOL).
FIRERS AND SCORERS, MOVE DOWNRANGE AND CHECK YOUR TARGETS.
MARK AND COVER ALL HOLES.

Table IV--Day Standing

E-25. The tower operator orders the firers to secure their weapons. On command, the scorer issues to the firer one magazine containing five rounds. The tower operator commands—

> TABLE FOUR, STANDING POSITION, 5 ROUNDS.
> LOAD AND CONDUCT STATUS CHECK (HOLSTER PISTOL).
> READY ON THE RIGHT.
> READY ON THE LEFT.
> READY ON THE FIRING LINE.
> MAKE READY (UNHOLSTER PISTOL).
> FIRERS, WATCH YOUR LANES.

E-26. The tower operator exposes the targets to the firers. When all targets have been exposed and engaged or lowered, the tower operator commands—

> CEASE FIRE.
> UNLOAD AND SHOW CLEAR (HOLSTER PISTOL).
> CLEAR ON THE RIGHT.
> CLEAR ON THE LEFT.
> THE FIRING LINE IS CLEAR.
> FIRERS AND SCORERS, MOVE DOWNRANGE AND CHECK YOUR TARGETS.
> MARK AND COVER ALL HOLES.

Table V--Day Moving Out

E-27. The tower operator orders the firers to secure their weapons and move to the center of the trail 10 meters behind the firing line. On command, the scorer issues to the firer one magazine containing one round; a second magazine containing seven rounds; and a third magazine containing five rounds. The tower operator commands—

> TABLE FIVE, STANDING POSITION, THIRTEEN ROUNDS.
> LOAD AND CONDUCT STATUS CHECK ONE MAGAZINE WITH 1 ROUND (HOLSTER PISTOL).
> LOAD YOUR SEVEN AND 5-ROUND MAGAZINES AT MY COMMAND.
> READY ON THE RIGHT.
> READY ON THE LEFT.
> READY ON THE FIRING LINE.
> PISTOLS AT THE READY POSITION.
> MAKE READY (UNHOLSTER PISTOL).
> FIRERS, WATCH YOUR LANES.
> MOVE OUT.

E-28. The tower operator exposes the targets to the firers. After each target or group of targets has been engaged, the tower operator commands—

> WEAPONS AT THE READY POSITION.
> MOVE OUT.

Qualification

E-29. After the firers complete Table V, the tower operator commands—
CEASE FIRE.
UNLOAD AND SHOW CLEAR (HOLSTER PISTOL).
CLEAR ON THE RIGHT.
CLEAR ON THE LEFT.
THE FIRING LINE IS CLEAR.
FIRERS AND SCORERS, MOVE DOWNRANGE AND CHECK YOUR TARGETS.
MARK AND COVER ALL HOLES.

E-30. The tower operator has each scorer total the firer's scorecard and turn it in to the range officer or the range officer's representative. The firing orders are rotated and the above sequence continued until all orders have fired.

Table VI--Day Standing, CBRN

E-31. The firer will wear a protective mask with hood. On command, the scorer issues to the firer one magazine containing seven rounds. The firer must get three hits to receive a GO on this table. The tower operator commands—
TABLE SIX, CBRN FIRE, STANDING POSITION, 7 ROUNDS.
LOAD AND CONDUCT STATUS CHECK (HOLSTER PISTOL).
READY ON THE RIGHT.
READY ON THE LEFT.
READY ON THE FIRING LINE.
MAKE READY (UNHOLSTER PISTOL).
FIRERS, WATCH YOUR LANES.

E-32. The tower operator exposes the targets to the firers. When all targets have been exposed and engaged or lowered, the tower operator commands—
CEASE FIRE.
UNLOAD AND SHOW CLEAR (HOLSTER PISTOL).
CLEAR ON THE RIGHT.
CLEAR ON THE LEFT.
THE FIRING LINE IS CLEAR.
FIRERS AND SCORERS, MOVE DOWNRANGE AND CHECK YOUR TARGETS.
MARK AND COVER ALL HOLES.

Table VII--Night Standing

E-33. The tower operator orders the firers to position. On command, the scorer issues to the firer one magazine containing five rounds. The firer must get two hits to receive a GO on this table. The tower operator commands—

TABLE SEVEN, NIGHT FIRE, STANDING POSITION, 5 ROUNDS.
LOAD AND CONDUCT STATUS CHECK (HOLSTER PISTOL).
READY ON THE RIGHT.
READY ON THE LEFT.
READY ON THE FIRING LINE.
MAKE READY (UNHOLSTER PISTOL).
FIRERS, WATCH YOUR LANES.

E-34. The tower operator exposes the targets to the firers. When all targets have been exposed and engaged or lowered. The tower operator commands—

CEASE FIRE.
UNLOAD AND SHOW CLEAR. (HOLSTER PISTOL).
CLEAR ON THE RIGHT.
CLEAR ON THE LEFT.
THE FIRING LINE IS CLEAR.
FIRERS AND SCORERS, MOVE DOWNRANGE AND CHECK YOUR TARGETS.
MARK AND COVER ALL HOLES.

ALIBIS

E-35. There are no alibis for the pistol qualification.

RULES

E-36. Certain rules apply to the conduct of fire during the CPQC. Many of those rules are described below for clarification.

Assistance

E-37. During instructional fire, the coach and assistant instructors should assist the firer in correcting errors. However, during record fire, no one may help or try to help the firer while or after the firer takes his or her position at the firing point.

Accidental Discharges

E-38. An accidental discharge will result in the termination of the Soldiers qualification and will be removed from the firing lane.

Fire on the Wrong Target

E-39. Each firer observes the location of the target in their own lane. Shots fired on the wrong target count as a miss. A firer is credited only for the targets hit in his or her own firing lane.

Qualification

Fire After the Signal to Lower Targets

E-40. Any shot after the target starts to lower is scored as a miss.

Extra Shot Fired at an E-Type Silhouette Target

E-41. If the firer hits the target while the target is exposed, that is, before it begins to lower, then the firer receives credit for the hit. The number of rounds fired to obtain the hit does not matter.

Excess Ammunition

E-42. At the end of each firing table, the firer turns in any excess ammunition. This ammunition is not re-issued to him or her for use in the other firing tables.

Target Sequence

E-43. The tower operator sets a common target sequence for all lanes. This keeps a firer from getting ahead of the firers in adjacent lanes. Target sequence varies in distance from the firer. It starts with 31 meters and allows for no more than two seven-meter targets.

SCORECARD

E-44. Figure E-1 on page E-10 shows a completed example of DA Form 88, *Combat Pistol Qualification Course Scorecard*. The blank form may be downloaded from the Army Publishing Directorate website. The scorecard lists the standards and provides scoring grids for the CPQC.

Appendix E

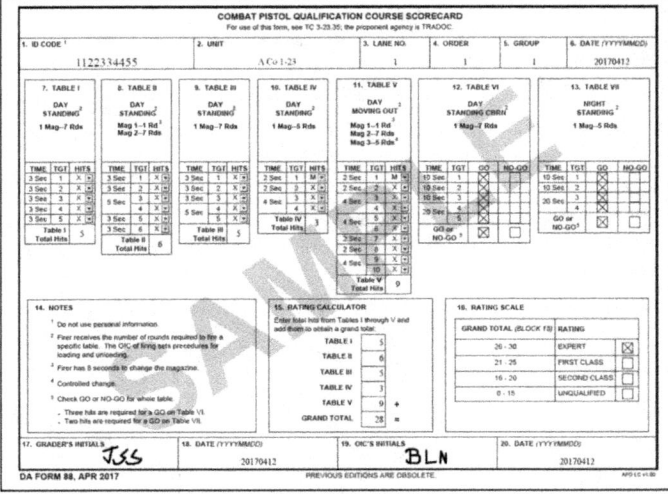

Figure E-1. Sample completed DA Form 88

TARGETS

E-45. Each firing lane requires seven electrical, device-type targets as well as a single E-type silhouette. Aggressor figures may be superimposed on the silhouettes to add realism to the course of fire.

QUICK-FIRE TARGET TRAINING DEVICE

E-46. The unit can get a quick-fire target-training device locally. To ensure standardization, quality, durability, and appearance, the device should be constructed by a qualified organization with documented experience producing similar devices such as the training aids section of the local Training Support Center.

Glossary

The glossary lists acronyms and terms with Army or joint definitions. Where Army and joint definitions differ, (Army) precedes the definition. Terms for which TC 3-23.35 is the proponent are marked with an asterisk. The proponent manual for other terms is listed in parentheses after the definition.

SECTION I – ACRONYMS AND ABBREVIATION

CBRN	chemical, biological, radiological, and nuclear
CPQC	combat pistol qualification course
ILL	illumination
IR	infrared
LOS	line of sight
NVD	night vision device
VIS	visible

SECTION II – TERMS

No terms defined by this manual.

This page intentionally left blank.

References

REQUIRED PUBLICATIONS

ADRP 1-02, *Terms and Military Symbols*, 16 November 2016.

DOD Dictionary of Military and Associated Terms, March 2017.

RELATED PUBLICATIONS

All websites are active as of 27 April 2017. Most Army doctrinal publications and regulations are available at: http://www.apd.army.mil.

Most joint publications are available online at: http://www.dtic.mil/doctrine/doctrine/doctrine.htm.

Other publications are available on the Central Army Registry on the Army Training Network, https://atiam.train.army.mil.

FM 27-10, *The Law of Land Warfare*, 18 July 1956.

TC 25-8, *Training Ranges*, 22 July 2016.

TM 9-1005-317-10, *Operator Manual Pistol, Semiautomatic, 9mm, M9 (1005-01-118-2640) (EIC: 4mn) Pistol, Semiautomatic, 9mm, M9A1 (1005-01-525-7966) Pistol, Semiautomatic, 9mm, Go Pistol (1005-01-588-5964) Air Force Only Go Pistol (1005-01-480-1274)*, 20 July 2016.

TM 9-5855-1911-13&P, *Operator and Field Maintenance Manual Including Repair Parts and Special Tools for Integrated White Laser Pointer (ILWLP), AN/PEQ-14 (Black) NSN 0855-01-538-0191 (Tan) NSN 5855-01-571-1258*, 31 July 2013.

TM 43-0001-27, *Army Ammunition Data Sheets for Small Caliber Ammunition (Federal Supply Class 1305) (Reprinted W/Basic INCL C1-13)*, 29 April 1994.

RECOMMENDED READING

AR 190-11, *Physical Security of Arms, Ammunition, and Explosives*, 5 September 2013.

ATP 3-21.8, *Infantry Platoon and Squad*, 23 August 2016.

TM 9-1005-317-23&P, *Unit and Direct Support Maintenance Manual (Including Repair Parts and Special Tools List) for Pistol, Semiautomatic, 9mm, M9 (NSN 1005-01-118-2640)(EIC: 4mn) and Pistol, Semiautomatic, 9mm, M9A1 (NSN 1005-01-525-7966) {TM 08993-IN/2; SW 370-AA-MMO-010/9mm; TO 11W3-3-5-4; Comdt Inst M8370.6} (This Item Is Included On EM 0065)*, 28 February 2007.

PRESCRIBED FORMS

Unless otherwise indicated, DA forms are available on the APD website (http://www.apd.army.mil).

DA Form 88, *Combat Pistol Qualification Course Scorecard.*
DA Form 5704, *Alternate Pistol Qualification Course Scorecard.*

REFERENCED FORMS

Unless otherwise indicated, (DA forms are available on the APD website (http://www.apd.army.mil).

DA Form 2028, *Recommended Changes to Publications and Blank Forms.*

WEBSITES

APD website: http://www.apd.army.mil/default.aspx
Comprehensive Soldier and Family Fitness (CSF2) Program: http://www.csf2.army.mil

Index

Entries are by paragraph number.

A
accuracy. 5-9
AN/PEQ-14. 3-3

C
cartridge case. A-2
chambering. 2-12
chemical operations. C-14
cocking. 2-18
components. 2-5
cycle of feeding. 2-11
cycle of function. 2-8

D
draw. 5-19

E
ejecting. 2-17
extracting. 2-16

F
firing. 2-14
firing hand. 6-6
firing positions. 6-33

G
grip. 6-5

H
holsters. 4-1

I
IR aim laser. 3-16
IR illuminator. 3-13
iron sight. 3-1

L
laser activation switch. 3-7
leg position. 6-2
locking. 2-13
lockout. 6-23

M
MAIN, 3-6
modes of operation. 3-4
muscle relaxation. 6-8

N
night vision devices. 3-3
nonfiring hand. 6-7

O
oblique targets. C-11

P
pattern generators. 3-27
point of aim. 7-20, 6-9
precision. 5-11, 5-9
primer. A-10
propellant. A-6

R
receiver. 2-7
recoil. 6-11
remedial action. 8-15

S
shot process. 7-9
sight alignment. 7-17
slide assembly. 2-6
stance. 6-3

T
target detection. 5-14
technical aspects. 2-1
toggle switch. 3-9
trapping method. C-9
turning movement. 9-8

U
unlocking. 2-15

V
visible aim laser. 3-19

W
weapon orientation. 7-10
white light illuminator. 3-23
workspace. 8-5

This page intentionally left blank.

TC 3-23.35
30 May 2017

By Order of the Secretary of the Army:

MARK A. MILLEY
General, United States Army
Chief of Staff

Official:

GERALD B. O'KEEFE
Administrative Assistant to the
Secretary of the Army
1714506

DISTRIBUTION:

Active Army, Army National Guard, and U.S. Army Reserve: To be distributed in accordance with the initial distribution number 110200, requirements for TC 3-23.35.

PIN: 201943-000

www.ingramcontent.com/pod-product-compliance
Lightning Source LLC
Chambersburg PA
CBHW050103230526
45470CB00004B/1659